FEMTOCELLS
OPPORTUNITIES AND CHALLENGES
FOR BUSINESS AND TECHNOLOGY

FEMTOCELLS
OPPORTUNITIES AND CHALLENGES
FOR BUSINESS AND TECHNOLOGY

By

Prof. Simon R. Saunders (editor)

Founding Chairman – Femto Forum, Independent Wireless Technologist and Visiting Professor – University of Surrey

Stuart Carlaw

Vice President, Mobile Wireless – ABI Research

Dr. Andrea Giustina

Vice President, Systems Engineering – Ubiquisys Ltd

Ravi Raj Bhat

Engineering Director – Continuous Computing

V. Srinivasa Rao

Architect for Wireless & IMS Trillium Protocol Software – Continuous Computing

Rasa Siegberg

Senior Systems Engineer, Mobile & Networking Security Solutions – SafeNet Inc.

A John Wiley and Sons, Ltd., Publication

This edition first published 2009
© 2009 John Wiley & Sons Ltd.

Registered office
John Wiley & Sons Ltd, The Atrium, Southern Gate, Chichester, West Sussex, PO19 8SQ, United Kingdom

For details of our global editorial offices, for customer services and for information about how to apply for
permission to reuse the copyright material in this book please see our website at www.wiley.com.

Library of Congress Cataloging-in-Publication Data

Femtocells : opportunities and challenges for business and technology / by Simon R. Saunders, editor ;
Stuart Carlaw ... [et al.].
      p. cm.
   Includes bibliographical references and index.
   ISBN 978-0-470-74816-9 (pbk.)
   1. Home computer networks–Equipment and supplies.   2. Wireless LANs–Equipment and supplies.   3. Cellular
telephone systems–Equipment and supplies.   4. Routers (Computer networks)   5. Radio relay systems.   6. Local
area networks industry.   7. Cellular telephone equipment industry.   I. Saunders, Simon R.   II. Carlaw, Stuart.
   TK5105.75.F46 2009
   621.382′1–dc22   2009007693

A catalogue record for this book is available from the British Library.

ISBN 9780470748169 (PBK)

Typeset in 10/12pt Times by Aptara Inc., New Delhi, India
Printed in Great Britain by CP1 Antony Rowe, Chippenham, Wiltshire

For
Gráinne, Luke and Emily.

Simon Saunders

For
Nandita and Aayush.

Ravi Raj Bhat

For
Rama Rao, Veerraghava, Sushma, Suhas.

V. Srinivasa Rao

Contents

About the Authors

Prof. Simon Saunders
PhD, CEng, FIET

Founding Chairman of the Femto Forum, Independent Wireless Technologist and Visiting Professor, University of Surrey.

Professor Simon Saunders is an independent specialist in the technology of wireless communications, with a technical and commercial background derived from senior appointments in both industry (including Philips and Motorola) and academia (University of Surrey).

He is an author of over 150 articles, books and book chapters. He has acted as a consultant to companies including BAA, BBC, O2, Ofcom, BT, ntl, Mitsubishi and British Land and was CTO of Red-M and CEO of Cellular Design Services Ltd. Simon speaks at and chairs a wide range of international conferences and training courses and has invented several novel wireless technologies. Particular expertise includes in-building wireless systems, radiowave propagation prediction, smart antenna design and mobile system analysis. He currently serves on technical advisory boards of several companies. He is the founding chairman of the Femto Forum, a Visiting Professor to the University of Surrey and is a member of the Ofcom Spectrum Advisory Board. See www.simonsaunders.com for further details.

Stuart Carlaw
Vice President Mobile Wireless – ABI Research.

Stuart Carlaw holds a program management role for ABI Research's Wireless Practice, which includes coordinating and planning its wireless infrastructure, mobile operator, wireless handset semiconductor, short range wireless connectivity, fixed mobile convergence, femtocell, mobile device and business mobility research services. He is also lead analyst for the femtocell, fixed-mobile convergence and mobile devices research services.

He brings to the task a deep understanding of the critical issues and technologies, stemming from his immersion in these markets for over a decade as both an analyst and an engineer.

Prior to joining ABI Research, Stuart was a Telecommunications Technician in the British Army, and then Senior Analyst and later Wireless Research Director at IMS Research. At IMS he was the primary Bluetooth analyst, and responsible for all short range wireless research.

He has been quoted and published in a variety of industry and general media, including BBC TV, BBC Radio 5, the *Financial Times*, the *Wall Street Journal*, *EE Times* and many more, and has spoken, moderated or participated in industry events including WiMAX World, TETRA World Congress, Bluetooth SIG AHM, PTT World Congress and Freescale Technology Forum.

Stuart earned a BSc (Honours) in Physical Sciences and Business Management from De Montfort University.

Dr Andrea Giustina
Vice President, Systems Engineering – Ubiquisys Ltd.

Dr Giustina joined Ubiquisys in early 2006 where he heads the definition of Femtocell offers and system design. With over 15 years of experience in telecoms, he previously held management positions in Lucent Technologies in Systems Engineering and Professional Services for GPRS, WCDMA and IMS Services and was a Researcher for Telecom Italia.

Dr Giustina holds a Doctor's degree in Electronics Engineering and a Telecommunications Master, both from Politecnico di Torino, and an MBA from Bath University.

Ravi Raj Bhat
Ravi Raj Bhat is an Engineering Director at Continuous Computing, managing its software R&D organisation. Ravi has published over 25 technical papers and articles in technical magazines including *IEEE Communication* and *Electronic Design*. Ravi had also co-authored a book titled, *Java in Telecommunication*, published by John Wiley & Sons in September 2001. Ravi holds an MBA from the Anderson School of Management at University of California, Los Angeles and BTech in Computer Engineering from National Institute of Technology, Karnataka.

V. Srinivasa Rao
V. Srinivasa Rao is an Architect for Wireless & IMS Trillium Protocol Software at Continuous Computing. In this capacity, Srinivasa is responsible for conceptualising initial product architecture and design. His oversight includes all the wireless and IMS protocol software products developed by CCPU. Srinivasa has published articles in technical magazines including *Electronic Design*. His expertise of communications systems ranges from the end-to-end network to signalling protocol software, with a primary focus on IMS and 3G wireless. Mr Srinivasa holds an MTech in Telecommunication Systems & Engineering from the Indian Institute of Technology, Kharagpur.

Rasa Siegberg

Senior Systems Engineer, Mobile & Networking Security Solutions, SafeNet Inc.

Rasa Siegberg joined SafeNet in 2005 as Systems Engineer responsible for technical presales and sales support, and various aspects of product management and marketing for the company's portfolio of embedded software security solutions.

Currently Rasa provides strategic guidance; sales, product management, and marketing support for all embedded hardware and software security products at SafeNet with a special focus on new technologies within the telecom/mobile industry.

He has more than 10 years of experience in telecom, semiconductor, networking and embedded security industries. Prior to joining SafeNet, Rasa worked as a Senior Consultant at Conformiq Software, and as a Product Manager at SSH Communications Security Ltd. He received his MA from Helsinki University, Finland.

Foreword

Inside Out

In many countries there are more (mobile) phones than people – it would be very easy to suggest that this means 'job done' or 'mission accomplished'. The reality is that there remain many opportunities and usage scenarios still to be reached by the evolution of mobile telephony in a digital broadband world. This is particularly the case when it comes to the twin requirements of in-building coverage and data services – the very heart of this text on femtocells edited and written by Simon Saunders and his colleagues.

Mobile was originally designed for mobility and particularly for outdoor coverage. The GSM and 3G digital standards were also designed for evolution, or natural selection as Darwin may have put it. The operator viewpoint has always been to maximise the shared R&D across many international markets in order to minimise costs and support the widest range of services and applications. The final frontier has always been seen as data and applications. However, without generous wireless spectrum and strong radio engineering these aspirations are not fully realised or implemented.

Typically the planning for domestic or in-building coverage has also been generally lower priority in the last 25 years of mobile communications. This relates both to spectrum available to operators, relative costs of in-building wireless solutions, and also to the availability of telecommunications alternatives. In the developed world, wireline communications have generally led the in-building communications connectivity, firstly with narrowband voice and more recently with broadband. This is about to change with the capability of bringing wireline and wireless technologies together – the breakthrough combinations of wireline backhaul with wireless femtocells promise so much more in terms of spectrum utilisation, capacity and quality, cost reduction and service differentiation.

This book covers these issues comprehensively with a great introduction to the role of femtocells and their growing capability. The Editor has provided a much-needed introductory text for students, engineers and practitioners alike to help grow this industry from the four billion plus mobile customers globally in the great outdoors, to the growing number of users and applications in the even greater indoor market. If we get this implementation right the prospect of Applications, Anywhere, Anytime can now fully be realised and this book will go a long way to help pragmatically in both understanding and meeting this vision..... inside out.

Dr Mike Short FBCS, FIET, FRGS
Vice President – R&D, Telefónica O$_2$ Europe
Former Chairman – Global GSM Association
Visiting Professor – University of Surrey

Preface

"Small things have a way of overmastering the great."
—Sonya Levien

I have been an enthusiast for in-building wireless systems for the last 14 years or so. They just make sense to me: why would you not put capacity and coverage where the people are? Why would you not deliver low power, high performance, spectrally efficient services, rather than building yet more expensive metal work on hilltops and rooftops? There are also lots of interesting technical challenges, which have been relatively neglected in both academic and industrial contexts.

At times in-building wireless has seemed like a hobby when compared to the serious business of outdoor systems. When I conducted research work into radio architectures and radiowave propagation for high capacity, high bit-rate indoor systems in the mid-1990s, funding bodies and colleagues questioned how such levels of capacity could ever be necessary – surely that many people making phone calls couldn't fit into a given room? Why would people need data at bit rates beyond a few tens of kilobits per second on a fixed line, let alone on a mobile phone?

The world looks very different now and the need for dedicated in-building systems increasingly seems like part of the prevailing orthodoxy. Wireless LAN systems exploded into the public domain a few years ago, playing a role in homes and in businesses, which few had predicted, and they became a standard part of enterprise IT equipment. Mobile systems increasingly included in-building systems for the largest buildings via picocells and distributed antenna systems. Yet the mobile architectures were typically hand-designed by relatively few, highly skilled individuals, making it difficult for them to scale to small businesses. The idea that mobile systems could deliver cost-effective services from inside the home was simply incredible.

So when I first heard about femtocells in 2005, it felt like coming across the missing link in an evolution which few were even aware of. It validated the technical potential – and the commercial impetus – which had seemed intuitively right to me a decade previously. I was immediately enthusiastic to get involved in some way. Early in 2007 I had that opportunity and worked with some of the pioneers in the field to found and chair the Femto Forum, an independent not-for-profit organisation of operators and technology firms working together to deliver femtocells.

I hope it is clear from this background that I am enthusiastic about femtocells, not because I chair the Femto Forum, but vice versa. I believe they are the right technology, at the right time, to meet the right needs of mobile users – which increasingly means the needs of just about everyone. There are plenty of challenges ahead in enabling widespread adoption of femtocells, but I think great progress towards this has been made in a remarkably short time and the signs for the future are mainly positive.

I've been asked many times in the last couple of years for good reference sources on femtocells. Apart from a few blogs, company promotional materials and articles in the press – which vary dramatically in their accuracy – there has been little to point them to. So I'm pleased to have worked together with an outstanding team of femtocell enthusiasts, each expert in their own field, to assemble this collection of writings on the many and varied aspects of femtocells. We think this should provide a sound basis for those seeking a rapid introduction to femtocells from a business or technical perspective. It should also provide a reasonable snapshot of the status of the femtocell market and standards. It is inevitable, however, that in a rapidly moving industry some of the content will be eclipsed by events. We would welcome comments on content which should be updated at the e-mail address below and will endeavour to address any issues in subsequent editions of the book. Errata and additional useful resources will be maintained at the website shown below.

Simon Saunders
For comments and suggestions: info@femtocellbook.com
For further information and useful resources: www.femtocellbook.com

Acknowledgements

I have had the privilege to work with some truly inspiring people over the last couple of years within the Femto Forum, who have broadened my exposure to the mobile world well beyond the specifics of femtocells. These are far too numerous to name. However, I must acknowledge Will Franks, Pete Keevill and Len Schuch for drawing the potential for femtocells to my attention in the first place. Many other individuals have provided great assistance in ideas, material and review, in particular Rupert Baines, Andy Tiller, Doug Knisely, Doug Pulley, Chris Smart, Alan Carter, Vicki Griffiths, Asan Khan, John Cullen, William Webb and Heather Kirksey. I am very grateful to Mike Short for his excellent foreword.

Icons in many of the figures are used by permission of Femto Forum Ltd.

Please note that the opinions expressed in this book are those of the individual authors and do not necessarily express the views of their companies or of the other authors.

Simon Saunders

The FAP data model illustrated in Chapter 6 is based on pioneering work done by Taka Yoshizawa, Michel Renaldo and Ian MacPherson in the Femto Forum. Their work has contributed significantly in moving femtocells closer towards a true 'plug and play' technology.

Ravi Raj Bhat, V. Srinivasa Rao

Abbreviations

1xRTT	1 times Radio Transmission Technology
2G	Second Generation Mobile
3G	Third Generation Mobile
3GPP	3rd Generation Partnership Project
3GPP2	3rd Generation Partnership Project 2
4G	Fourth Generation Mobile
A	BSC–MSC interface (GSM)
A5/3	The Kasumi block cipher algorithm
AAA	Authentication, Authorisation and Accounting
Abis	BTS–BSC interface (GSM)
ACL	Access Control List – list of users who can access femtocell service from a FAP
ACS	Auto Configuration Server – device providing connectivity between CPEs and network operators' OSS to configure and administer CPE devices
ADSL	Asymmetric Digital Subscriber Line
AES	Advanced Encryption Standard – a symmetric crypto algorithm
AGC	Automatic Gain Control
AH	Authentication Header
AP	Access Point
ARPU	Average Revenue Per User
AS	Application Server
ASN	Access Service Network
ASN GW	ASN Gateway
ASN.1	Abstract Syntax Notation One – a standard notation used to describe data structures for encoding, transmitting over network and decoding
BBF	Broadband Forum – previously known as DSL Forum, it is a consortium of vendors and operators dedicated to developing specifications to accelerate development and deployment of broadband networks
BS	Base Station
BSC	Base Station Controller
BTS	Base Station Transceiver System
CA	Certificate Authority
CAPEX	CAPital EXPenditure

CC	Call Control
CDMA	Code Division Multiple Access
cdma2000	Family of mobile standards standardised by 3GPP2 and comprising CDMA2000 1xRTT, CDMA2000 EV-DO and CDMA2000 EV-DV
CENELEC	French: Comité Européen de Normalisation Electrotechnique. English: European Committee for Electrotechnical Standardization
CM	Cable Modem – CPE providing broadband access over cable television infrastructure
CMTS	Cable Modem Termination System – an equipment in cable service providers' networks, which terminates cable modem connections to provide broadband services
CN	Core Network
CPE	Customer Premises Equipment – a device in an end-user environment providing access to a service provider's network
CPU	Central Processing Unit
CS	Circuit Switched
CSG	Closed Subscriber Group
CSN	Core Service Network
CVC	Code-Verification-Certificate
CWMP	CPE WAN Management Protocol – an application layer protocol specified by Broadband Forum in TR-069 for remote management of CPE
DAS	Distributed Antenna System
DECT	Digital Enhanced Cordless Telecommunications
DHCP	Dynamic Host Control Protocol – used by IP clients to resolve hostname, IP address and other information in an IP network
DNS	Domain Name Server – Internet server to translate hostname to IP addresses
DOCSIS	Data Over Cable Service Interface Specification – an international standard which defines the communication and operation support interface requirements for a data over cable system
DoS	Denial of Service – a form of network attack
DSM-CC	Digital Storage Media Command and Control
DSP	Digital Signal Processing
E911	Enhanced 9-1-1 – emergency calling
EAP	Extended Authentication Protocol
EAP-AKA	Extensible Authentication Protocol for UMTS Authentication and Key Agreement
EAP-SIM	Extensible Authentication Protocol SIM
EIRP	Effective or Equivalent Isotropic Radiated Power
EM	Electro-Magnetic
EPC	Evolved Packet Core
ESP	Encapsulating Security Payload – one of the IPsec protocols
ESS	Enhanced System Selection
ETSI	European Telecommunications Standards Institute
EV-DO	Evolution – Data Optimised or Only
EVRC	Enhanced Variable Rate Codec

Fa	FAP–FGW interface
FAP	Femtocell Access Point – the femtocell CPE
FAP	Femto Access Point
FAP-MS	FAP Management System
Fa-R6	FAP–ASN GW interface
Fas	IMS CN–Femto AS interface
Fb	FGW–CN interface
Fb-cs	FGW–CS CN interface
Fb-ims	FGW–IMS CN interface
Fb-ps	FGW–PS CN interface
FCAPS	Fault, Configuration, Accounting & Administration, Performance and Security – set of services required to manage communication network
FCC	Federal Communications Commission
FDD	Frequency Division Duplex – a telecommunications duplexing method where the transmitter and receiver at each terminal use different carrier frequencies
Fg	FGW–FGW-MS interface
FGW	Femto Gateway
FGW-MS	FGW Management System
FLUTE	File Delivery over Unidirectional Transport (multicast protocol)
FM	Frequency Modulation
Fm	FAP–FAP-MS interface
FMC	Fixed Mobile Convergence
FMS	Femto Management System; Fixed-Mobile Substitution
FPGA	Field Programmable Gate Array
FTP	File Transfer Protocol
FTTx	Any variant of the Fibre To The: Home, Building, Cabinet, etc...
Fx1	FAP–MGW interface (CDMA)
Fx2	FAP–IMS interface (CDMA)
Fx3	FAP–FGW interface (CDMA)
Gb	BSC–SGSN interface
GGSN	Gateway GPRS Support Node
GPRS	General Packet Radio Services
GPS	Global Positioning System – technology to identify the precise location of an object via precise microwave signals using a constellation of low earth-orbiting satellites
GSM	Global System for Mobile Communication – 2G wireless technology defined by European Telecommunications Standards Institute (ETSI) and now maintained by 3GPP
GTP-U	GPRS Tunnelling Protocol User data
GW	Gateway
HNB	Home Node B
HeNB	Home evolved Node B
HeNB GW	HeNB Gateway
HeNB MS	HeNB Management System

HFC	Hybrid Fibre/Coaxial – access network of optical fibre and coaxial cable in cable modem network
HGW	Home Gateway
HMS	HNB Management System
HNB	Home Node B
HNB GW	HNB Gateway
HPLMN	Home PLMN
HRPD	High Rate Packet Data (commonly known as 1xEV-DO)
HSDPA	High Speed Downlink Packet Access – member of the HSPA family for mobile broadband access in downlink (RAN to mobile device) direction, defined by 3GPP
HSPA	High Speed Packet Access – a family of mobile broadband technology as overlay over UMTS network
HSPA+	Evolved HSPA as defined in 3GPP Release 7 and beyond
HSUPA	High Speed Uplink Packet Access – member of HSPA family for mobile broadband access in uplink (mobile device to RAN) direction
HTTP	Hypertext Transport Protocol – used for retrieving interlinked text documents
HTTPS	Hypertext Transfer Protocol Secure
HUK	Hardware Unique Key – an unalterable and physically protected private key for a given device
ICNIRP	International Committee on Non-Ionising Radiation Protection
ICS	IMS Centralized Services
ID	Identifier
IEEE 802.11a/b/g/n	IEEE standards for wireless LAN, on which Wi-Fi is based
IEEE 802.16	IEEE standards for wireless MAN, on which WiMAX is based
IETF	Internet Engineering Task Force
IKE	Internet Key Exchange – protocol used to set up a security association
IKEv2	Internet Key Exchange (v2) – protocol for authentication, key generation and session set up; used in combination with the IPsec protocol suite
IKEv2	Internet Key Exchange version 2
IMS	IP Multimedia Subsystem
IMSI	International Mobile Subscriber Identity
IMT-2000	International Mobile Telecommunications-2000 (IMT-2000) is the global family of standards for 3G wireless communications as defined by the ITU
IMT-Advanced	ITU global family of standards for wireless communications beyond IMT-2000, sometimes referred to as 4G
IP	Internet Protocol
IP TV	IP Television
IPDR	IP Detail Record – provides information about IP-based service usage and other activities for OSS
IP-PBX	IP – Private Branch eXchange
IPR	Intellectual Property Rights
IPv4/v6	Internet Protocol version 4/ version 6

IPsec	IP security – a suite of protocols for providing security (privacy and integrity) for IP protocol
ISO	International Organization for Standardization – an international standard setting body composed of representatives from various national standards organisations
ISP	Internet Service Provider
IT	Information Technology
ITU	International Telecommunication Union
ITU-R	ITU Radiocommunication Sector – one of three sectors of the ITU
Iu	RNC–MSC/SGSN interface in 3GPP
Iub	NodeB–RNC interface in 3GPP
Iuh	HNB–HNB GW interface in 3GPP
Iu-UP	RNC–MGW interface
L1	Layer 1
L2	Layer 2
LAN	Local Area Network
LI	Lawful Intercept
LIA	Local IP Access – means of bypassing wireless core network, while on a femtocell, to get access to IP content, thereby conserving precious wireless core network resources; also known as *data offload* and *local breakout*
LOS	Line-Of-Sight
LTE	Long-Term Evolution
M3UA	MTP level 3 User Adaptation
MAC	Medium or Media Access Control
MAP	Mobile Application Part
MFIF	MAP-Femto Interworking Function
MGW	Media Gateway
MIB	Management Information Base – a collection of data used to manage devices in a communication network
MIMO	Multiple Input Multiple Output – space time coding technology using multiple antennas at base stations and mobiles
MIPS	Million Instructions Per Second
MMS	Multimedia Messaging Service
MM	Mobility Management
MMTel	Multimedia Telephony
MNO	Mobile Network Operator
MO	Managed Object – an abstract network resource to be managed; an element in a MIB
MS	Mobile System; Mobile Station
MSC	Mobile Switching Centre
MVNO	Mobile Virtual Network Operator
NGMN	Next Generation Mobile Network – mobile networks beyond 3G
NGN	Next Generation Network – architectural evolutions of telecommunication core and access networks

NMS	Network Management System – system used by service provider to administer/manage a network
NodeB	UMTS base station
Nonce	Number used only once
NTP	Network Time Protocol – means of transmitting time signals over a communication network
O&M	Operations & Maintenance
OAM&P	Operation, Administration, Maintenance and Provisioning
OCXO	Oven Controlled Crystal Oscillator
OECD	Organisation for Economic Cooperation and Development
Ofcom	Office of Communications – the United Kingdom communications regulator
OFDM	Orthogonal Frequency Division Multiplexing
OFDMA	Orthogonal Frequency Division Multiple Access
OID	Object Identifier – an identifier used to uniquely identify a managed object in a MIB
OPEX	Operational Expenditure
OSI	Open Systems Interconnection – an abstract description for layered communications and computer network protocol design
OSS	Operational Support System – network used by communication service provider for FCAPS service
PAD	Peer Authorisation Database – provides mapping between security association management protocol and policy to be applied (from SPD) to the security association
PBX	Private Branch eXchange
P-CPICH	Primary Common Pilot Channel
PDCP	Packet Data Convergence Protocol
PDU	Protocol Data Unit – an abstract set of bytes used by protocols to communicate over a network
PHY	PHYsical layer
PKI	Public Key Infrastructure
PLMN	Public Land Mobile Network – established and operated by network service providers for the specific purpose of providing land mobile communication services to the public
PN	Pseudo-Noise or Pseudo-random Number sequence
PS	Packet Switched
PSTN	Public Switched Telephone Network
PTP	Precision Time Protocol
PUZL	Preferred Use Zone List
QoS	Quality of Service
R1	UE–BS interface (WiMAX)
R3	ASN–CSN interface
R4	ASN–ASN interface
RAB	Radio Access Bearer – set of services provided by a Radio Access Network to transfer user data between mobile device and core network

RAN	Radio Access Network – network of elements (such as Node Bs and Radio Network Controllers) that enable mobile devices to access core network services
RAN 1/2/3/4	TSG RAN WG 1/2/3/4 – working Groups of the RAN technical specification group of 3GPP
RANAP	RAN Application Protocol
RAT	Radio Access Technology
REM	Radio Environment Measurement – set of information gathered to evaluate the quality of radio signals
RF	Radio Frequency
RFC	Request For Comments
RLC	Radio Link Control
RNC	Radio Network Controller – a network element in UTRAN that controls Node Bs
RNC	RAN Network Controller
RPC	Remote Procedure Call – inter-process communication mechanism that allows a process to call subroutine for execution in a process running in another address space
RRC	Radio Resource Control
RSC	Radio Spectrum Committee
RSCP	Received Signal Code Power
RTCP	RTP Control Protocol – out-of-band control information for RTP flow
RTP	Real-time Transport Protocol – defines a standardised packet format to deliver audio and video over the Internet
RTP	Real Time Protocol
RTWP	Received Total Wideband Power
RUA	RANAP User Adaptation
S1-MME	eNB–EPC control interface
S1-U	eNB–EPC user plane interface
SA	Security Association – establishment of shared security information between two network entities to support secure communication
SA1/5	TSG SA WG 1/5 – working groups of the Service Aspects TSG in 3GPP
SAD	Security Association Database – stores parameters associated with each security association
SAR	Specific Absorption Rate
SCCP	Signalling Connection Control Part
SCTP	Stream Control Transmission Protocol
SeGW/SecGW	Security Gateway
SFTP	Secure File Transfer Protocol
SGSN	Serving GPRS Support Node
SI	French: Le Système International d'Unités; English: The international system of units
SIM	Subscriber Identity Module
SIP	Session Initiation Protocol
SIP UA	SIP User Agent
SME	Small/Medium Enterprise

SMI	Structure of Management Information – a subset of ASN.1 used in SNMP to define sets of related managed objects in a MIB
SMS	Short Message Service
SNMP	Simple Network Management Protocol – an application layer protocol to monitor network attached devices
SNOW	A block cipher
SOAP	Simple Object Access Protocol – used for exchanging structured information to implement web services
SOHO	Small Office/Home Office
SON	Self-Organising Network(s)
SPD	Security Policy Database – set of policies that applies to all IP traffic for secure communication
SSID	Service Set IDentifier
SSL	Secure Sockets Layer – cryptographic protocol that provides security and data integrity for communications over the Internet
Syslog	Standard for forwarding log messages in an IP network
TAS	Telephony Application Server
TCP	Transmission Control Protocol
TCXO	Temperature Controlled Crystal Oscillator
TDD	Time Domain Duplex – a telecommunications duplex indexing method where transmitter and receiver take turns on same carrier frequency
TD-LTE	Time Division LTE
TEE	Trusted Execution Environment – a secure subsystem within a device for security sensitive tasks
TFTP	Trivial File Transfer Protocol – used to transfer files from one networked device to another
TLS	Transport Layer Security – cryptographic protocol that provides security and data integrity for communications over the Internet
TMN	Telecommunications Management Network – a protocol model defined by ITU-T for managing open systems in a communication network
TMSI	Temporary Mobile Subscriber Identity
TR	Technical Requirement
TrE	Trusted Environment (see TEE)
TS	Technical Specification
TSG	Technical Specification Group
TSG-A	3GPP2 TSG: Access Network Interfaces
TSG-C	3GPP2 TSG: Radio Access
TSG-S	3GPP2 TSG: Service and System Aspects
TSG-X	3GPP2 TSG-X: Core Networks
UDP	User Datagram Protocol – transport protocol used in the Internet for un-assured data transfer services
UE	User Equipment (handset, data terminal or other device)
UICC	Universal Integrated Circuit Card – used in mobile terminals to ensure integrity and security of data stored on mobile terminals
Um	MS–BTS interface

UMTS	Universal Mobile Telecommunication System – a 3G wireless technology with uses WCDMA as its underlying air interface
URL	Uniform Resource Locator
USIM	Universal Subscriber Identity Module
UTRAN	UMTS Terrestrial Radio Access Network – a collective term for NodeBs and Radio Network Controllers in UMTS Radio Access Network
Uu	UE–NodeB interface
UWB	Ultra WideBand
VoIP	Voice over IP
VPN	Virtual Private Network – virtual network (usually secure) overlayed over physical network for secure communication among group of users
WAN	Wide Area Network – a data-communication network that covers a broad area (i.e., any network whose communications links cross metropolitan, regional or national boundaries)
WCDMA	Wideband CDMA
WG	Working Group
WHO	World Health Organisation
Wi-Fi	Wireless Fidelity
WiMAX	Worldwide interoperability for Microwave Access
X.509	An ITU standard for a public key infrastructure
xDSL	Any variant of the DSL technology: DSL, ADSL, VDSL, etc.
XML	eXtensible Mark-up Language

List of Figures

List of Tables

1

Introduction to Femtocells

Simon Saunders

1.1 Introduction

In this chapter we establish the basic 'why, what and how' for femtocells. All of the issues discussed here are covered in greater depth in later chapters, but this chapter should serve as a rapid introduction to the whole subject.

1.2 Why Femtocells? The Market Context

Mobile phones have been one of the fastest-growing consumer technologies of all time. Digital mobile phones were introduced in the early 1990s, and have now grown to include around 4 billion mobile phone subscriptions worldwide – nearly 60% of the world's population. The number is continuing to grow quickly, and is expected to reach 5.63 billion by 2013 (1). Mobile phone data traffic is forecast to grow by between 10 and 30 times between 2008 and 2013, depending on the pricing and promotion of these services (2).

In the same period, the Internet has also become a mass-market technology, growing to 1.6 billion users worldwide, nearly 25% of the world's population (3). Internet protocol traffic is forecast to grow by over 10 times in the period from 2006 to 2012 (4).

Since the introduction of third-generation mobile services in the early part of the new millennium, the dream has been to combine mobile and Internet technologies, giving fast, reliable access to the Internet via personal mobile devices. While there have been false starts in achieving this dream, there are now clear signs that demand for Internet services is taking off. In 2007 particularly, the availability of 3G networks, usable mobile devices and flat-rate near-unlimited data plans came together to produce a tipping point in the take-up of mobile data services. This rapid growth has been exhibited in terms of both the quantity of mobile broadband data consumed and the revenues derived from this data, and this growth is widely forecast to continue and even accelerate (5).

Revenues generated by many mobile operators from mobile broadband data also increased substantially in 2007, as shown in Table 1.1. Yet the table also shows that voice revenues are

Femtocells: Opportunities and Challenges for Business and Technology   Simon R. Saunders, Stuart Carlaw, Andrea Giustina,
Ravi Raj Bhat, V. Srinivasa Rao and Rasa Siegberg   © 2009 John Wiley & Sons, Ltd

Table 1.1 Growth of operator revenues for leading operators in Q3 2007, relative to the previous 12-month period (6)

Mobile operator	Data revenue growth	Voice revenue growth
AT&T	64%	6%
Verizon Wireless	63%	7%
Rogers	53%	15%
Telstra	50%	5%
Vodafone (W. Europe)	45%	1%
Sprint	28%	–9%
T-Mobile Germany	24%	–4%
KDDI	18%	1%

growing much more slowly and are declining in some cases, as prices fall under competitive and regulatory pressures.

Although the growth of data in 2007 was from a relatively small base, the combination of this with the relatively flat market for mobile voice services (at least in developed economies) has made mobile data a significant and growing component of overall operator revenues, reaching nearly 20% in 2007 (6). The overall market is already very large. For example, the financial analysts Merrill Lynch had this to say in 2007 (6) (our emphasis):

> Wireless data services are now a $115bn global market, growing at 28% annually, contributing ~2 pts to aggregate telecom services revenue growth – *outstripping fixed broadband revenues and growth.*

So operators have a compelling reason to pay close attention to the mobile broadband data market. Yet they also have challenges. While both data volumes and revenues have increased, the volumes have increased far faster than the revenues and this trend is expected to accelerate in the future (Figure 1.1). In order to maintain healthy margins, operators need to find ways to substantially decrease the cost per bit of delivering this data, while not placing limits on customers' appetites for consuming the data.

1.3 The Nature of Mobile Broadband Demand

To determine how to serve mobile broadband demand cost effectively, it is important to understand the nature of this demand – and most particularly, where it occurs.

Traditionally, mobile service was primarily about services to mobile *users* – those travelling between homes and offices via public or private transport, where communication services are not available in any other way.

Increasingly, however, the most important – and biggest - components of mobile demand are services delivered to mobile *devices*, yet stationary *users*. In other words, mobile devices are increasingly personal and are used by individuals when they are not on the move, typically within buildings, including both the home and the workplace.

As a reasonable rule of thumb, roughly one-third of all cellular traffic today is at home despite networks not typically being designed to provide a solid home service and tariffs often being unattractive compared with fixed-line networks. Another third is in the workplace, with the remaining third being the 'traditional' traffic generated on the move. It is expected that

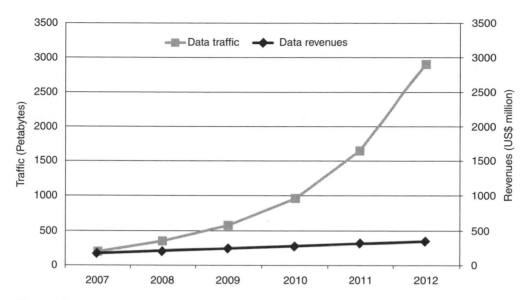

Figure 1.1 Forecast global mobile data revenues and traffic. *Reproduced by permission of Informa Telecoms and Media Ltd*

these proportions will grow in the future even as the overall data volumes grow, due to the need for users to be looking at a screen when consuming most high-bandwidth data services (e.g. web browsing or mobile video) – see Figure 1.2.

Studies of these traffic patterns show that in Western Europe 57% of mobile minutes at home or work (8). In-building traffic on 3G networks is expected to grow to 75% of the total by 2011 (9). Home coverage remains patchy: for example, the UK is generally considered a well-developed market with over 90% population coverage for 3G and far higher for 2G (10). Nevertheless, a detailed study of users in the UK showed that 19% of mobile phone owners regularly encounter coverage problems in the home (11). Of these, 53% report that coverage is poor throughout the house, while the remainder report coverage problems in selected rooms.

Why is this growth happening? We can imagine numerous influencing factors, including the following:

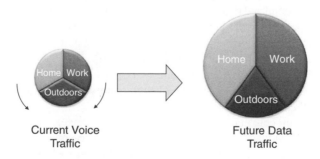

Figure 1.2 Increasing proportions of traffic generated indoors

- Service take-up often starts at home and then spreads to the enterprise: witness the way in which Wi-Fi, although originally intended as an enterprise technology, only reached a mass market via usage in homes before being adopted in corporate environments.
- Voice minutes are increasingly moving from fixed line to mobile, as an increasing proportion of users come to rely upon their mobile as their main – or even only – telephone.
- Operators (not only mobile ones) are increasingly offering 'quadplay' and other bundled flat-rate tariffs, which include large quantities of mobile service along with fixed line, Internet and television services.

There are increasingly real, compelling data services which users wish to use on mobiles:

- mobile multimedia – including television;
- media synchronisation (wireless sideloading);
- presence applications;
- consumer push email.

For more information on the potential evolution of future mobile applications, see this author's predictions in (12).

Any and all of these services point towards the need for operators to deliver a high-quality service to users at home and at work, including reliable coverage and high data rates, while reducing the cost per bit for delivery. Femtocells have emerged at just the right time to address these issues.

1.4 What is a Femtocell?

A femtocell is a low-power access point, based on mobile technology, providing wireless voice and broadband services to customers in the home or office environment. As shown in Figure 1.3, the femtocell connects to the mobile operator's network via a standard consumer broadband connection, including ADSL, cable or fibre. Data to and from the femtocell is carried over the Internet – or at least, over an Internet-technology network provided by an Internet service provider.

Typically, a single femtocell will deliver voice services simultaneously to at least four users within the home, while allowing many more to be connected or 'attached' to the cell, accessing services such as text. Additionally, femtocells will deliver data services to multiple users, typically at the full peak rate supported by the relevant air interface technology, currently several megabits per second and rising to tens and hundreds of megabits per second in the future.

Data from multiple femtocells are concentrated together in a gateway, managed by the mobile operator, and ultimately find their way back to the operator core network along with the data from the conventional operator macrocell network. The operator core network also contains a management system which provides services to the femtocell, ensuring that the services experienced by the user are secure, of high quality and can coexist with the signals from other femtocells and the outdoor network.

In practice, the femtocell may be either a stand-alone device, which connects into the customer's existing broadband router, or may form a key component of a home gateway device which incorporates the router and other technologies, such as a broadband modem, Internet router and Wi-Fi access point into a single integrated device. Examples of both types are illustrated in Figure 1.4. Note that femtocells are consumer devices, intended to be suitable

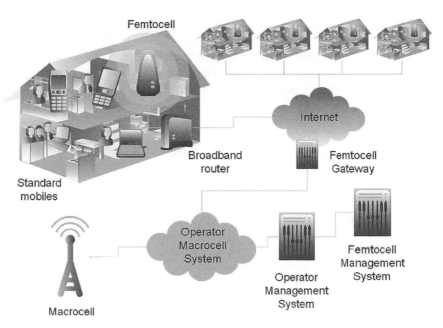

Figure 1.3 Basic femtocell network. *Reproduced by permission of Femto Forum Ltd*

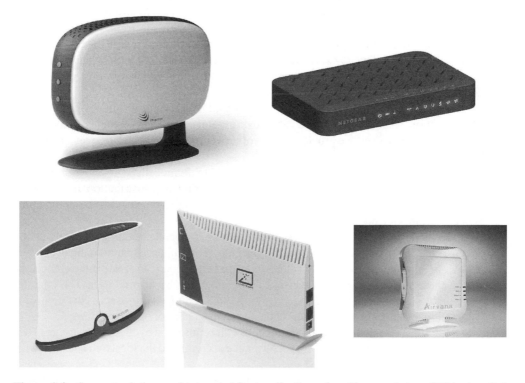

Figure 1.4 Some stand-alone and integrated femtocells. *Reproduced by permission of Ubiquisys Ltd., ip.access Ltd., Radio Frame Networks Inc. and Airvana Inc*

for installation in a home environment and to be manufactured in large volumes in line with other consumer products.

In some respects, femtocells resemble other home wireless devices such as cordless phones and Wi-Fi access points. However, there are important differences, which are highlighted in Section 1.6. A clear definition of a femtocell is therefore required.

1.4.1 Femtocell Attributes

The Femto Forum has created the following set of attributes, *all* of which are necessary for a device to qualify as a femtocell.

A femtocell is a low-power wireless access point, incorporating all of the following:

- **Uses mobile technology**. Femtocells use fully standard wireless protocols over the air to communicate with standard mobile devices, including mobile phones and a wide range of other mobile-enabled devices. Qualifying standard protocols include GSM, WCDMA, LTE, Mobile WiMAX, CDMA and other current and future protocols standardised by 3GPP, 3GPP2 and the IEEE/WiMAX Forum, which collectively comprise the technologies included in the ITU-R definition of IMT.[1] The use of such protocols allows femtocells to provide services to several billion existing mobile devices worldwide and to provide services that users can access from almost any location as part of a wide-area network.
- **Operates in licensed spectrum**. By operating in spectrum licensed to the service provider, femtocells allow operators to provide assured quality of service to customers over the air, free from harmful interference but making efficient use of their spectrum.
- **Generates coverage and capacity.** As well as improving coverage within the home, femto-cells also create extra network capacity, serving a greater number of users with high data-rate services. They differ in this from simple repeaters or 'boosters' which may only enhance the coverage.
- **Over Internet-grade backhaul**. Femtocells backhaul their data over Internet-grade broad-band connections, including DSL and cable, using standard Internet protocols. This may be over a specific Internet service provider's network, over the Internet itself or over a dedicated link.
- **Permits low prices.** The large volumes envisaged for femtocells will allow substantial economies of scale, driving efficiencies in manufacturing and distribution in a manner similar to the consumer electronics industry and with pricing projected to be comparable with access points for other wireless technologies.
- **Fully managed by licensed operators.** Femtocells only operate within parameters set by the licensed operator. While they have a high degree of intelligence to automatically ensure that they operate at power levels and frequencies that are unlikely to create interference, the limits on these parameters are always set by operators, not by the end user. The operator is always able to create or deny service to individual femtocells or users. This control is maintained whether the femtocell itself is owned by the operator or the end user.
- **Self-organising and self-managing.** Femtocells can be installed by the end customer. They set themselves up to operate with high performance according to the local and network-wide

---

[1] International Mobile Telecommunications, comprising IMT-2000 (usually known as 3G) and IMT-Advanced (which may become known as 4G).

conditions regarding radio, regulatory and operator policies, with no need for intervention by the customer or operator. They continue to adjust themselves over time as the customer, operator and regulator needs evolve to maximise performance and reliability.

1.4.2 Femtocell Standards

Most air interfaces included in the global ITU-R IMT family have active programmes to develop standards for femtocells. These include:

- 3GPP standards for Home Node-B, which is a WCDMA femtocell. Both FDD and TDD options are likely and a TD-SCDMA variant is also planned.
- 3GPP standards for Home eNode-B, which is an LTE femtocell. Both FDD and TDD options are envisaged.
- 3GPP2's programme for femtocells for cdma2000, cdma2000 1x, HRPD, 1x EV-DO and UMB.
- WiMAX Forum's programme for WiMAX femtocells based on IEEE standards.

In all cases femtocell standards will support deployments in all of the existing licensed spectrum bands in which macrocells operate.

1.4.3 Types of Femtocell

Individual femtocells are likely to come in various hardware types. Although individual standards differ in their definitions, the following broad classes can be identified, though these are not exclusive or prescriptive:

- **Class 1.** This is the class of femtocells that has emerged first and is currently best known. Femtocells in this class deliver a similar transmit power and deployment view to Wi-Fi access points (e.g. typically 20 dBm of radiated power[2] or less) for residential or enterprise application. They will each deliver typically 4–8 simultaneous voice channels plus data services, supporting closed or open access. Installed by the end-user.
- **Class 2.** Somewhat higher power (typically up to 24 dBm of radiated power), perhaps to support longer range or more users (say 8–16). Supports closed or open access. May be installed by the end-user or the operator. May be viewed as an evolution of picocell technology.
- **Class 3.** Still higher power for longer range or more users (e.g. 16 or greater). Typically carrier deployed and may well be open access. Could be deployed indoors (e.g. in public buildings) for localised capacity, outdoors in built-up areas to deliver distributed capacity or in rural areas for specific coverage needs.

1.5 Applications for Femtocells

Femtocells started as a means of delivering services to residential environments. This remains a core application for femtocells and it enables femtocell technology to be produced in large

---

[2] Effective Isotropic Radiated Power - EIRP

volumes and low costs. However, femtocells are not limited to this application and early deployments for other purposes are anticipated. Current applications include:

- **Residential** – Femtocells are installed indoors within the home by the end user and may be stand-alone devices or integrated with other technology such as residential gateways. Access to the residential femtocell will often be closed – restricted to a specified group of users – but may also be open to all registered users in some cases. Typically these application needs will be met using class 1 femtocells.
- **Enterprise** – Enterprise femtocell deployments may be in small-office, home-office situations, in branch offices or in large enterprise buildings. Femtocells for this purpose are usually of class 1 or class 2 and will typically support additional functionality compared with residential devices such as handover between femtocells, integration with PBX and local call routing. Will primarily be used indoors, but could also be used to serve a corporate campus. Installation will probably be managed by the carrier, but may be achieved by the enterprise itself or its IT subcontractors. Access may be closed or open.
- **Operator** – This class encompasses a wide variety of applications where operators use femtocells to solve specific coverage, capacity or service issues in both indoor and outdoor environments. These could be composed of class 1, 2 or 3 devices and will usually be open access. They will be installed by the operator or by third parties under the operator's direction.
- **Others** – These application classes are not exclusive and it is expected that other innovative ideas for the application of femtocells will emerge, for example on aircraft, trains or passenger ferries. In all cases the essential attributes of femtocells described earlier will be observed, enabling full compliance with relevant local customer, operator and regulatory requirements.

1.6 What a Femtocell is not

A femtocell could be confused with other devices. It is helpful to contrast its behaviour with several of these, specifically Wi-Fi access points, repeaters (or boosters), cordless telephones and conventional cellular base stations.

Wi-Fi Access Points While these also provide wireless broadband access to portable devices, there are important differences. Wi-Fi almost always operates in unlicensed (or licence-exempt) spectrum. This means that an operator is unable to guarantee any service quality over the air, since interfering devices can legally appear close to any given user. Most Wi-Fi devices operate in the 2.4 GHz frequency band, where only three non-overlapping channels are available, so the potential to avoid interference is limited. In contrast, femtocells, operating in licensed spectrum, ensure that an operator is in control of every transmitting device and can manage interference to deliver an appropriate quality of service to every user. Wi-Fi access points and client devices all transmit at a power of around 100 mW, which does not change even when far less power is required, increasing the incidence of interference and draining batteries. By contrast, in cellular technology both the mobile devices and the femtocells continually adjust their transmit power to the minimum necessary to deliver adequate service, increasing the number of users which can access a given channel, reducing interference and increasing battery life.

Wi-Fi is mainly associated with delivering Internet or private LAN connectivity to laptop devices. In this role it has been extremely successful, with the majority of new laptops including embedded Wi-Fi capability. In contrast, while femtocells can serve the increasing numbers of laptops with cellular connectivity (either embedded or via 'dongles'), they are designed primarily to serve the much larger numbers of personal devices including mobile phones and newer ultra-mobile PCs.

Relatively recently, Wi-Fi has been extended to serve 'dual-mode' devices such as phones. However, the number of such devices is tiny compared with cellular phones, so while there may be a market for such devices it is not expected to address the huge potential user base addressable by femtocells.

Early Wi-Fi failed to provide adequate data security. While these issues have been addressed in later standards, the issue remains in many legacy devices and causes considerable concern and confusion for users. By contrast, the security standards used in cellular systems have been proven to be robust, are well trusted by users, and are used directly in femtocells, together with additional safeguards to ensure the authentication and encryption of devices and traffic connected across the Internet.

For the avoidance of doubt, then, there is no such thing as a 'Wi-Fi femtocell'.

Repeaters (or 'Boosters'). Repeaters are bi-directional amplifiers, used to increase coverage in systems, including cellular mobile. They operate by receiving service from an outdoor base station cell via an external antenna, amplifying it, and retransmitting it via cables and an antenna placed within the area for coverage improvement. They are a useful tool in many networks, and can certainly improve coverage, but there are many differences from femtocells. Repeaters require careful professional installation for good results, with the external and internal antennas requiring proper location to deliver appropriate coverage while being isolated from each other to avoid feedback. The gain of the amplifier must be adjusted within tightly defined limits: too little will fail to deliver adequate coverage improvement, while too much will cause feedback between the antennas, degrading or even denying service for all the users in the coverage of the external 'donor' cell. By contrast, femtocells are zero-touch devices, requiring no professional skills for set-up. As well as amplifying the signals from the mobile devices, repeaters necessarily inject some noise into the outdoor base station receiver. This limits the number of repeaters which may be used in a given cell, typically to a handful per cell, since more will unacceptably degrade the base station performance for all other users in a cell. Femtocells, with appropriate interference mitigation techniques, may be used in very large numbers in a given area. Repeaters require that a given location already has acceptable outdoor coverage and only requires improvement indoors, while femtocells can operate in a completely isolated area provided broadband connectivity is available. Although repeaters can deliver the improved coverage, which is one important motivation for femtocells, they do not deliver additional capacity into the network or any capability to provide differentiated services. Lastly, repeaters tend to be far more expensive than the target prices for femtocells, even before the cost of professional services is counted.

Cordless Telephone Systems Cordless phones and their associated base units, originally based on analogue FM and now typically based on DECT, were one of the first wireless devices to be deployed in volume in the home. Their success lies in delivering convenience to users, of being able to access fixed lines while being able to move freely around the home. In that respect they are similar to femtocells. However, they are single-function devices, with few delivering anything other than voice. They are not personal devices, in that handsets are shared amongst all users. They do not deliver data services in the main. Most importantly, they deliver no wide-area mobility (although DECT-GSM dual-mode devices were once available). It is entirely possible that cordless phones will diminish in popularity once femtocells are widely available.

Cellular Base Stations Femtocells do share much in common with conventional mobile base stations, producing almost indistinguishable signals over the air in order that standard handsets can be used unmodified. However, there are many differences. Femtocells have limited capacity, suitable for a single domestic installation, while conventional base stations must serve tens or hundreds of users. Base stations are therefore professional products, with costs to match, while femtocells are consumer products, produced in volumes to meet consumer pricing expectations. Femtocells are designed to work over Internet-grade backhaul, typically DSL or cable, while base stations operate over dedicated backhaul, such as leased lines or microwave links. As a result, femtocells do not have the same interfaces as standard base stations: the interface is optimised to reduce the bandwidth requirements, while also increasing the level of security to protect traffic which may be routed over the Internet rather than a dedicated network. In order to provide local management of system and radio resources, they include much more intelligence than a conventional base station, being comparable to the combination of a base station and a 'collapsed' radio network controller, which would usually manage the resources of several base stations together. Base stations would usually be planned and optimised by professional radio planners, while such intelligence is automatic within femtocells. Lastly, femtocells of course radiate at substantially lower powers than base stations, being typically around 10 000 times lower. So while a femtocell does have some similarities with a base station, it is both much more and much less than one, and the two should not be confused.

These differences are summarised in Table 1.2. Overall, while femtocells have much in common with some other devices, drawing on proven technologies and customer demand where appropriate, they represent a unique new class of device.

1.7 The Importance of 'Zero-Touch'

It is vital for femtocells to be simple to install, configure and operate. No special expertise must be necessary, since the cost of sending trained personnel to consumers' homes would destroy the business case for the operator. Further, for users to adopt femtocells, femtocells must deliver service with an absolute minimum of intervention by the user. There should be no need to provide special security settings or to access any device (e.g. a computer) other than the femtocell and the user's mobile device. This 'zero-touch' aspect of femtocells is critical for any femtocell deployment and is illustrated by Figure 1.5, which shows a typical extract from a femtocell user manual. Note the 10-minute configuration period. During this time the femtocell is engaged in some very sophisticated activities, including:

- Contacting the mobile network and establishing a secure, fully encrypted communications tunnel.

Table 1.2　Comparison of femtocells and other wireless devices

	Wi-Fi access points	Cordless telephones	Repeaters ('boosters')	Cellular base stations	Femtocells
Operates in licensed spectrum	✗	✗	✓	✓	✓
Supports power control	✗	✗	✓	✓	✓
Robust security	✗	✓	Not applicable	✓	✓
Serves existing personal devices	✗	✗	✓	✓	✓
Provides voice and data services	✓ (but requires device-specific extensions)	✗	✓	✓	✓
Provides wide-area mobility	✗ (unless combined with mobile systems and special enhancements)	✗	Not applicable	✓	✓
Supports existing personal devices	✗	✗	✓	✓	✓
Consumer device and price	✓	✓	✗	✗	✓
Consumer installation	✓	✓	✗	✗	✓

- Mutually authenticating with the network and establishing the user's credentials, service options and location.
- Surveying the surrounding radio environment and configuring itself to operate within parameters set by the operator to deliver good service with minimal interference.

Users are entirely unaware of this activity other than a delay of a few minutes. They should simply be able to start accessing their femtocell services at the end of the period. They can subsequently leave the femtocell switched on permanently, or power it down when not required for an extended period. The femtocell will provide service much more quickly on subsequent start-up.

1.8 User Benefits

So why would a mobile user wish to use a femtocell? It is essential that the answer to this question is well articulated and developed if femtocells are to be successful. There are many parts to the answer, and the most appropriate will differ according to the market and market segment into which femtocells are sold. Operators will choose the most appropriate for the users to whom they are selling femtocells. A range of potential user benefits is summarised in Figure 1.6 and described individually here.

- *Coverage.* The most fundamental user benefit is the provision of reliable coverage throughout the home. This allows users to rely on their mobiles as a prime means of making and receiving

1 Insert the SIM card   **2** Plug in the data cable   **3** Plug the other end into your router/modem

4 Plug in the power   **5** Leave for 10 minutes   **6** When the green lights are constantly on it's ready

Figure 1.5   Zero-touch femtocell installation. *Reproduced by permission of Ubiquisys Ltd*

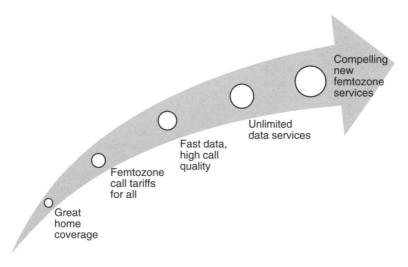

Compelling new femtozone services

Unlimited data services

Fast data, high call quality

Femtozone call tariffs for all

Great home coverage

Figure 1.6   The range of user benefits associated with femtocells

calls. Even where coverage is already nominally available in a home, currently users will often turn to a fixed line to avoid the risk of dropping a call and will typically suggest the use of fixed-line numbers to friends and business associates when an important call is expected. In many cases, home coverage is patchy, perhaps available only on upper floors or close to windows. In many markets, residential densities below a certain level will lead to no coverage at all due to the difficulty of operators providing coverage with an adequate business case using conventional solutions. In all of these cases, femtocells should result in mobile phones being able to be used in a better way, allowing users to rely on their mobiles at home as their prime – or only – personal communication device.

- *Femtozone call tariffs.* Some femtocell operators include special tariffs for calls made or received on the femtocell, including a lower rate, a large bundle of minutes per month or matching the fixed-line rates. A wide range of tailored pricing options is possible, differentiating by user type, time of day, call destination and other dimensions. Some of these benefits may also be available to users when granted access to femtocells from the same operator in their friends' or colleagues' homes.
- *Fast data and high call quality.* The improved coverage and protection from interference available from femtocells enables mobile phones to work at the peak of their capability, including the highest possible data rates available over the air and the highest possible call quality. This is likely to lead to users making new uses of their phones and other mobile-enabled devices, including relying on them for the most important business calls and using them as their primary personal Internet and entertainment devices.
- *Unlimited data services.* Operators are likely to offer data services on femtocells which have far higher usage limits or are even uncapped, enabling users to use these services without fear of the associated charges. This will also encourage users to use such services for the first time, overcoming initial barriers to access and ultimately to adopt these services beyond the home as well.
- *Compelling new femtozone services.* As well as delivering existing mobile network services better and at more attractive prices, femtocells can also deliver brand new services, which use the specific knowledge of the user's location to offer extra benefits such as control of devices around the home, synchronisation of content to and from the mobile device and fast, high-quality access to data stored on other devices on the home network such as media servers.

1.9 Operator Motivations and Economic Impact

The delivery of services via femtocells impacts on the economics of those services for operators in all major dimensions, including increasing revenue, decreasing costs and speeding up deployment.

Revenue impact:

- Femtocells deliver new revenue streams from value-added services and by increasing mobile usage, both within the home and on the wider network, as users demand services they have first experienced on the femtocell.
- Location-specific tariffs such as 'homezone' tariffs defined by the coverage area of a few macrocells may extend over a large area potentially comprising the whole of a small town and therefore lead to revenue leakage. This decreases operator revenues without decreasing

the associated costs. Location-specific tariffs delivered via femtocells are precisely targeted and reduce the costs of delivery, thereby increasing the overall value to the operator.

- Family and group contracts, which increase the loyalty of the femtocell users to the operator, extending both the value and lifetime of the contract. The attractiveness of such contracts is also increased by the ability to offer service bundles including entertainment services, which are typically bought by the whole family rather than by a single individual.
- Differentiated services, encouraging users to adopt services from operators offering femtocells over those available from other operators.

Cost savings:

- Femtocells enable operators to defer and reduce the cost of macrocell roll-out to deliver enhanced indoor coverage and network capacity.
- Femtocells produce operational savings – especially on the major items associated with sites in power, backhaul and site rental.
- By reducing the likelihood of customer churn between operators, femtocells reduce the cost of customer retention.

Time-to-market:

- Femtocells allow rapid, low-risk, focused deployment of next-generation technologies such as LTE and WiMAX, avoiding the extended timescales and cost needed to secure site rights and construction resources to permit macrocell upgrades.
- New services can be rapidly provisioned to relevant users and locations, enabling them to be trialled and optimised for wider roll-out more quickly and with lower risk, enabling Internet-style speeds of new service introduction.

Overall, there is potential for substantial economic value to be created by the deployment of femtocells, challenging the traditional preconceptions of cellular economics.

1.10 Operator Responses

Operators in many regions of the world have responded to the potential for femtocells with an enthusiastic and rapid response. The first commercial femtocell deployments took place in the Sprint network in the United States in 2007 based on CDMA technology and further commercial launches based on WCDMA started in late 2008 and early 2009, including Starhub in Singapore, Softbank Mobile in Japan and Verizon in the United States.[3] Many other operators around the world are working towards service launches and wider rollout within 2009 and 2010. Publicly announced femtocell trials include Vodafone, T-Mobile, AT&T, Telefonica O2, mobilkom austria, Verizon and TeliaSonera. Examples of comments from operators include the following:

- 'Our Apple iPhone and flat rate data tariffs place huge capacity demands on our networks. Because so much of that usage is at home, femtocells could play a crucial role in underpinning the explosive growth of mobile broadband usage' (Vivek Dev, COO, Telefonica O2 Europe).

---

[3] This list will certainly be incomplete by the time this book is published. See www.femtocellbook.com for a more current list.

- '3G femtocells answer the needs of our mobile-centric strategy 100%' (Frank Esser, CEO, SFR).
- 'We intend to use [femtocells] to reduce macro capex spend by up to 20 percent in some areas' (Andy MacLeod, Director of Group Networks, Vodafone).
- 'Femtocells are fundamental to the future of mobile. They pave the way for new mobile services that put the mobile phone at the centre of the connected home' (Axel Kolb, Fund Manager, T-Mobile Venture Fund).

1.11 Challenges

As we have seen, there are many reasons why customers and operators are demanding femtocells and the services and economies that can be provided by using them. Yet there are many challenges in achieving this potential. These are briefly highlighted here and examined in greater depth in later chapters.

Market challenges:

- public awareness;
- public concerns regarding service, tariffs and alleged health issues;
- support for a wide range of use cases;
- business case.

Radio and physical challenges

- interference management between femto- and macrocells;
- radio resource and mobility management;
- implementation issues (e.g. synchronisation, signal processing and cost).

Network challenges

- architectures and interfaces;
- management and provisioning;
- scalability;
- security.

Regulatory challenges

- regulatory benefits;
- spectrum issues;
- service issues.

These issues and others besides have required significant and swift attention to permit femtocells to realise their full potential.

1.12 Chapter Overview

The remainder of this book is organised as follows.

Chapter 2 examines the background for small cells, placing femtocells in the context of the history of other solutions for in-building service and explains the market and technological factors which have made femtocells a compelling proposition.

Chapter 3 addresses market issues, covering the benefits and motivations of femtocells for operators, the key market challenges, business case analysis and forecasts for the take-up of femtocells.

Chapter 4 looks at radio issues for femtocells, including the requirements and methods for interference management between femtocells, femtocell RF specifications and health issues.

Chapter 5 addresses the options for the network architecture of femtocells, particularly the interfaces and protocols for integrating femtocells with the operator network across the Internet and the way in which the various options differ between standards families.

Chapter 6 describes the requirements and approaches for provisioning and managing millions of femtocells in an efficient and scalable fashion.

Chapter 7 explains the security aspects of femtocells from a customer and operator perspective, including analysis of the potential threats and solutions.

Chapter 8 covers the standards for femtocells across the main mobile standards families, namely 3GPP, 3GPP2 and WiMAX Forum. It also introduces the industry bodies who are playing important roles in the introduction and proliferation of femtocells.

Chapter 9 shows the benefits of femtocells as seen from a regulatory perspective and highlights areas where regulations may need to be evolved to maximise these benefits and avoid delay to femtocell take-up.

Chapter 10 explains some of the considerations which are important for efficient implementation of femtocells in hardware and software.

Chapter 11 characterises the services that operators can offer to femtocell users, particularly the opportunity to extend services beyond those available from the macrocell network.

Finally Chapter 12 summarises the outcomes of this book and gives some predictions of the potential future for femtocells.

2

Small Cell Background and Success Factors

Simon Saunders

2.1 Introduction

This chapter explains the context for femtocells, including some of the historical background, the alternative small-cell solutions, which have been available for addressing indoor coverage and capacity needs,[1] and the unique combination of factors which make femtocells technically viable and commercially attractive at the present time.

2.2 Small Cell Motivations

2.2.1 Cellular Principles

First, we need to examine why small cells are increasingly needed in mobile systems.

The fundamental characteristic of cellular radio systems is that an unlimited amount of traffic can be served by a limited bandwidth of spectrum. This is achieved by reusing spectrum increasingly densely as the number of users in a given area increases, and by ensuring that the interference so generated is controlled and managed to avoid detriment to the services experienced by the users. This principle was recognised as early as 1947 in internal work at Bell Laboratories (13).

The most basic method of achieving this spectrum reuse is simply to directly reuse frequencies at base stations which are sufficiently separated from each other for the interference to be minimal. This gives rise to the traditional image of the cellular industry showing hexagons which approximate the coverage area of multiple base stations, with the same numbers/shading representing base stations which share the same frequencies (Figure 2.1).

---

[1] A video lecture covering similar ground is available at www.femtocellbook.com.

Femtocells: Opportunities and Challenges for Business and Technology   Simon R. Saunders, Stuart Carlaw, Andrea Giustina,
Ravi Raj Bhat, V. Srinivasa Rao and Rasa Siegberg   © 2009 John Wiley & Sons, Ltd

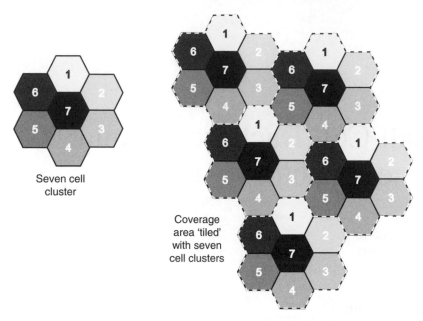

Seven cell
cluster

Coverage
area 'tiled'
with seven
cell clusters

Figure 2.1   Reuse of frequencies in an idealised cellular network to increase system capacity (after (14))

In practice, the methods used to reuse spectrum in modern systems are more subtle, and the real world is not populated with hexagonal coverage areas, but the principle still applies and remains extremely powerful. By allowing operators to plan and optimise the reuse of frequencies within their licensed band without recourse to a regulator, the traffic densities per unit bandwidth for mobile systems have become far higher than those for any other wireless system.

In order to increase the bit rate for mobile broadband services, and to maximise the capacity available from a given quantity of spectrum and number of cells, much research and development effort has been expended on improving the format of the signals transmitted and on the technology in the transmitters and receivers, allowing the spectrum efficiency of systems (in bits per second per hertz) to be increased substantially. Yet this alone is insufficient to meet the growing traffic demands. For example, the transition from third-generation to fourth-generation mobile systems is expected to increase spectrum efficiency by around two to five times, and has taken about a decade to achieve. While this is impressive, it does not come close to meeting customer demands, which are increasing by orders of magnitude in just a few years. While additional spectrum bands will be needed to meet the promise of these newer systems, small cells remain the only way of scaling up total system capacity sufficiently.

2.2.2 Conventional Cell Types: Why 'Femtocells'?

Cell types have evolved towards smaller cells over the years, even before the emergence of femtocells. The conventional hierarchy of cell types is illustrated in Figure 2.2 and explained below.

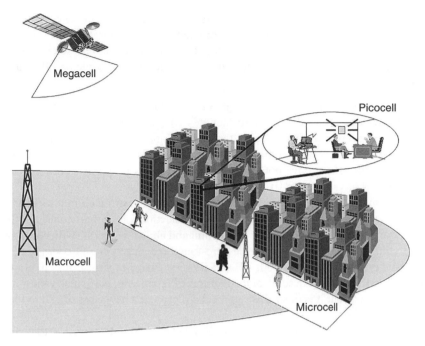

Figure 2.2 Hierarchy of conventional cell types, from (14). *Reproduced by permission of John Wiley & Sons Ltd*, © 2007

Megacells These are provided by satellite systems (or by high-altitude platforms such as aircraft or lighter-than-air ships (15)) to mobile users, allowing coverage of very wide areas with reasonably low user densities. A single satellite in a low earth orbit would typically cover a region of 1000 kilometres in diameter. Most systems operate at L and S bands (1–4 GHz) to provide voice and low-rate data services, but systems operating as high as Ka band (26–40 GHz) can be deployed to provide Internet access at high data rates over limited areas.

Macrocells Designed to provide mobile services (including both voice and data), particularly outdoors, to rural, suburban and urban environments with medium traffic densities. They are the dominant approach in today's mobile networks. Base station antenna heights are greater than the surrounding buildings, providing a cell radius from around 500 metres to many tens of kilometres. Antennas are usually mounted on towers or on building rooftops. Mostly operated at VHF and UHF (30 MHz–3 GHz). May also be used to provide fixed broadband access to buildings at high data rates, typically at UHF and low SHF frequencies (several GHz to tens of GHz).

Microcells Designed for high traffic densities in urban and suburban areas to users both outdoors and within buildings. Base station antennas are lower than nearby building rooftops, so coverage area is defined by the street layout. Cell length up to around 500 metres. Again mostly operated at VHF and UHF, but services as high as 60 GHz have been studied.

Picocells Very high traffic density or high data rate applications in indoor environments. Users may be both mobile and fixed; fixed users are exemplified by wireless local area networks (LANs) between computers. Coverage is defined by the shape and characteristics of rooms, and service quality is dictated by the presence of furniture and people and the surrounding interference.

So both microcells and picocells use prefixes borrowed from the prefixes used in specifying small quantities in SI units, where 'micro' denotes 10^{-6} and 'pico' denotes 10^{-9}. 'Nano' for 10^{-9} has been used as a product-specific term and is not in wide use. So 'femto', denoting 10^{-12}, has been introduced to convey clearly that femtocells are operating at an even smaller scale than picocells. There have not so far been any serious proposals for attocells (10^{-18}), zeptocells (10^{-21}) or yactocells (10^{-24}).

However, femtocells are distinguished from picocells not so much due to their size, but by virtue of the distinctive attributes highlighted in Section 1.4. In particular, femtocells are suitable for domestic use in terms of the price point and ability to operate with consumer-grade broadband connections. They also feature zero-touch management and configuration, while picocells have typically required both RF and IP networking expertise. It is likely that, over time, picocells will incorporate some of the features of femtocells, particularly the autonomous management aspects, but it is clear that femtocells represent a new class of cell to the existing hierarchy.

We now examine why the dominant macrocells need to be complemented by the introduction of more small cells.

2.2.3 Challenges of Achieving Indoor Coverage from Outdoor Macrocells

The main traditional method of achieving indoor coverage is from outdoor macrocells. Radio planning rules – link budgets – are developed which include an allowance for signal penetration into buildings, so that the density of macrocells is higher than would be required for outdoor coverage alone. The level of this allowance is dependent on the propagation characteristics of the buildings in the area and the level of reliability of service desired. This delivers some level of indoor coverage for large numbers of buildings economically and ensures wide area mobility. However, users now demand increasing reliability and use their mobiles for a greater proportion of their calls and for more demanding data applications. To meet these needs without sacrificing quality the required link budget allowance increases rapidly. The number of macrocells needed rises even faster and rapidly reaches an economic limit where the increased expenditure is not justified by increased revenue. Some areas can never be economically covered in this way, such as tunnels, basements and sparsely populated areas, where the number of dwellings is too low for a macrocell to make sense.

For example, Figure 2.3 shows the way in which the number of macrocells rises as the depth of indoor coverage is increased by a small distance. The specific number of macrocells required depends on the propagation conditions, but if we take the specific attenuation to be 1 dB per metre, then the number of macrocell sites has to increase by around 27% for just one metre of extra coverage depth. For an operator of a modest size with 5000 sites initially and an assumed cost (capex and opex) per site of 225,000 euro (net present value), the cost is 308 million euro for this extra metre. In practice the average cost per site may even increase for the

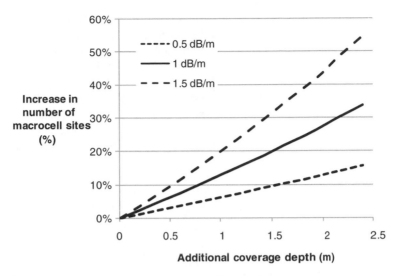

Figure 2.3   Range of increase in macrocell sites required to deliver additional indoor coverage depth, depending on the attenuation per unit distance inside the buildings. The outdoor path loss exponent is assumed to be 38 dB per decade of range (see Chapter 8 of (14) for details)

extra sites since suitable macrocell sites are becoming more costly – and time consuming – to acquire and construct and may be subject to increasingly restrictive planning regulations.

Further, if the demand for additional indoor coverage arises from the need for higher data rates and/or data capacity, the interference between sites becomes increasingly difficult to control as the site density increases.

Moves towards using lower frequency bands (e.g. the 700, 800, 850 and 900 MHz bands applicable in various world regions) for indoor coverage do help with providing indoor coverage economically. However, they tend to defer, rather than avoid, the point at which an alternative solution is needed and the proportion of sites to increase coverage depth still follows the trend indicated in Figure 2.3. They also do not help with delivering capacity or interference control, or indeed in the delivery of new tariffs or other new services which are dedicated to particular buildings.

Overall, this argues for an alternative solution that is complementary to macrocell coverage and which delivers indoor coverage from a solution dedicated to the purpose. Putting it another way, delivering indoor coverage exclusively from outdoors to indoors is inside-out thinking. Or even more strongly, providing indoor coverage from macrocells is like trying to fill a cup from a fire hose! However, the move to small cells requires that the costs scale downwards at least as fast as the number of users served reduces from the macrocell case in order to remain economical.

2.2.4 Spectrum Efficiency

The demand for additional data services from wireless systems is rising very rapidly, while the spectrum available to deliver it is not. The supply of spectrum is increasing to some extent, by virtue of operators opening up new frequency bands and removing restrictions on existing ones, but the overall quantity is not increasing sufficiently quickly for this to be the only solution. These new bands also tend to be at higher frequencies, where the costs of delivery are even higher and require costly replacement of mobile devices. So it is essential to make better use of existing spectrum, by increasing the overall spectrum efficiency of the networks which use it.

There are many ways of measuring spectrum efficiency, but if we consider the spectrum efficiency as a measure of the density of users who can simultaneously access a given data throughput, or the total throughput available per unit area. Then we can express the spectrum efficiency in bits per second per hertz per square kilometre or similar units as follows:

$$Spectrum\ efficiency = \frac{Total\ data\ throughout}{Spectrum\ utilised \times Unit\ area}$$

$$= \frac{Data\ throughout\ per\ cell \times Number\ of\ cells}{Spectrum\ utilised \times Unit\ area}$$

$$= \frac{Data\ throughout\ per\ cell}{Spectrum\ utilised} \times Cell\ density$$

$$= (Modulation\&\ Coding\ Efficiency) \times (Cell\ density)$$

This shows that there are two basic ways for the spectrum efficiency to be increased. We can either increase the *modulation and coding efficiency* of the system, i.e. its ability to deliver a given bit rate in a given spectrum in each cell, or we can increase the *cell density* in the system. Most of the attention to date has gone into increasing the modulation and coding efficiency and each generation of wireless technology has achieved impressive gains over the previous one. However, the opportunity for doing this is decreasing, rather than increasing over time. Unlike Moore's law, which is an empirical observation on the progress of semiconductor engineering (16), the bounds on the capacity of a given communications link are fixed by Shannon's law (17), which is rooted in inviolable statistics. The performance of modern wireless systems comes increasingly close to Shannon's limit, with little subsequent gain to be achieved. This is widely recognised, for example, the chief executive of Qualcomm was quoted as saying (18), 'the improvement of wireless links that enhance user throughputs is reaching its limit'.

For example, the target performance for the design of the LTE standard was to achieve an increase in spectrum efficiency per site (equivalent to modulation and coding efficiency) of three to four times that of HSDPA (downlink) and two to three times (uplink) depending on the conditions (19). Even if that is achieved, it does not come close to the levels of capacity which will be needed given most forecasts of the growth of mobile broadband services, e.g. (2).

This is consistent with historical observations, crystallised in 'Cooper's law' (20), which states that the total capacity available in a given area has approximately doubled every 30 months over at least the last 100 years, resulting in an increase of around one million times in the last 45 years. Of this remarkable increase, around 25 times came from the ability to access more of the available spectrum. Another 25 times came from the increase in the modulation efficiency, five times of which was due to using spectrum in frequency division and another five times from the increase via modulation schemes. The remaining 1600-times improvement came from increases in cell density.

So, to continue to sustain the growth in the use of wireless services, there is a clear requirement for an increase in the density of cells and for the cost of those cells to be reduced.

This creates a clear mandate for femtocell development. Again, the Qualcomm chief executive noted (18):

> According to the results of our research [into femtocells], this effort will possibly result in eight times higher throughput per user. In retrospect, an eight-times improvement is equivalent to that brought by the cell phone's shift from analog to digital.

2.2.5 Geometry Factors

As well as increasing the density of reuse of spectrum, small cells, deployed carefully, also increase the *geometry factor* of the system. The geometry factor expresses the ratio of the power available in the wanted cell to the interference power coming from surrounding cells. If the geometry factor stays constant, the relative level of the interference stays constant, and each cell can only achieve a constant total capacity for a given air-interface technology. However, by making positive use of the radio wave propagation characteristics of the environment, small cells can additionally increase the isolation between cells. For example, in microcells the isolation of the surrounding buildings contains the signal from each cell within its coverage area and prevents leakage to other cells. The same role is played by the building walls for femtocells and picocells, especially the outer wall of the building, which typically has the highest propagation loss and provides for isolation between the macrocells and the indoor cells. The small-cell users experience very high geometry factors over most of the cell area, which are only available in macrocells very close to the base station and thus for a small proportion of users. This increases the maximum data rate available to users and also allows the network to deliver a given quantity of data in less time, reducing the amount of interference in total in the system or increasing the total network capacity.

For example, Figure 2.4 shows how the throughput available to one user in an HSDPA system varies with the geometry factor under various conditions for HSDPA. For macrocells, the G-factor $\left(G = 10 \log \dfrac{Wanted\ cell\ power}{Other\ cell\ interference\ power} \right)$ is around −3 dB near the cell edge and has a median value (for 50% of users) of around 2 dB. In good conditions a macrocell user may achieve around 12–15 dB. Small-cell users will achieve G-factors near the top of

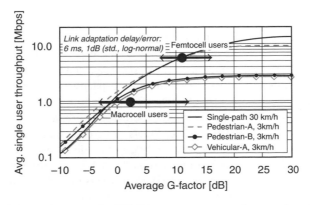

Figure 2.4 Throughput performance for HSDPA versus geometry factor, adapted from (22). *Reproduced by permission of John Wiley & Sons Ltd,* © 2004

this range over most of the cell. In practice the result for femtocells may be even better than this figure suggests due to the low levels of multipath delay in femtocell environments. See (21) for further discussion of these issues.

2.2.6 The Backhaul Challenge

Traditionally, the main limitation for operators to deliver large quantities and rates of data has been the radio interface. As radio interfaces increasingly come close to their theoretical limits, however, the challenges of delivering backhaul from cell sites are increasingly the main limitation. Traditional backhaul works over dedicated leased lines. Cell sites have often been installed with T1 or E1 lines delivering around 1.5 Mbps, which was ample for typical voice traffic, but can be exceeded by the demands of a single mobile broadband data user today. Increasingly the demand from next-generation networks is suggesting a need for backhaul of as much as 450 Mbps upstream and 150 Mbps downstream per site (23). While a variety of solutions are being examined to meet these needs, all come at substantially increased cost, which does not necessarily scale well with the revenue associated with these services.

By designing the interface between the cell and the operator's core network to be more bandwidth efficient and tolerant of contention with other users, operators have the opportunity to reuse backhaul intended for consumer applications. Moreover, customers will typically already be paying for this backhaul, so the operator economics are transformed. The usc of the Internet as a widely distributed backhaul and transport medium also reduces the impact of bottlenecks at concentration points and Internet protocols increase the overall network resilience.

2.3 Other Small-Cell Systems

2.3.1 Overview

Having established clearly that small cells are an important and increasingly necessary component of wireless networks, we must recognise that femtocells are not the only form of smaller cell which needs to be considered.

2.3.2 Picocells

Picocells have existed for a substantial period and arose from a recognition of the need for high-quality in-building coverage in particular locations, especially large office buildings. In such buildings, however, the demanding nature of the usage and the relatively large number of users meant that picocells needed essentially the same capacity as a macrocell and necessitated the use of dedicated backhaul, partly because standards and technology of the time did not support any other option. They also required expert installation and optimisation. The result was that the first wave of early picocell products were essentially low-power macrocells, delivering few cost savings. They were thus only deployed in the largest corporate accounts, where the revenues (or the avoidance of churn) made such expenditure worthwhile. In many cases, large infrastructure vendors dropped picocell products from their offerings, preferring to support macrocells with distributed antenna systems for such applications (see next section).

Latterly, a second wave of picocells based on IP backhaul has been successful in delivering 2G services in particular niches. These niches have included smaller enterprises, remote areas where operators do not have an existing macrocell network, business parks, underground locations and unusual locations such as on-board ships and on aeroplanes. Beyond these niches, however, the growth of picocells has been limited, partly due to the need for detailed operator involvement in their design and deployment. They do not address the needs of the home or home office environment or of the smallest office buildings.

For 3G services, however, the need is more pressing. If picocells can leverage the potential economies of scale associated with femtocells, while also delivering the operational efficiencies arising from the self-managing nature of femtocells, they could address a wider set of business applications in the future. Indeed, the third wave of picocells may essentially be an evolution from femtocells and could build substantially on the successes of both femtocells and of the second wave of picocells.

2.3.3 Distributed Antenna Systems

Distributed antenna systems are an elegant means of distributing signals from a single source to the required areas of coverage or capacity around a building. The source may be a base station dedicated to serving the building, or a repeater, extending coverage from outdoor macrocells via an external antenna. The distribution of signals may either be via a passive network (Figure 2.5), where unpowered RF splitters/combiners are used to separate the power around a network of coaxial cables and antennas, or via an active network where bi-directional amplifiers are inserted to overcome cable losses (Figure 2.6). Active systems can use copper cables or optical fibres.

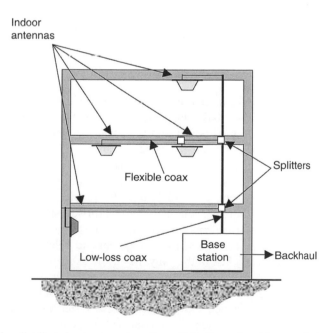

Figure 2.5 Passive distributed antenna system, from (14). *Reproduced by permission of John Wiley & Sons Ltd*, © 2007

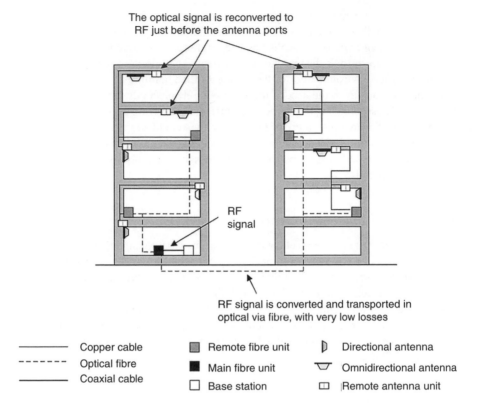

Figure 2.6   Active distributed antenna system, from (14). *Reproduced by permission of John Wiley & Sons Ltd*, © 2007

The basic advantage of a DAS is that appropriate placement of the antennas overcomes the propagation losses between the base station and the mobile users due to walls and floors of the building. As a result, the total power and the power per antenna can be substantially reduced compared to attempting to cover the same area using a single picocell or macrocell. Additionally, the channels in the signal source are available to all antennas, so less base station capacity is needed for a given grade of service (i.e. significant trunking gain is achieved). As a result, distributed antenna systems are widely used for providing services in large public buildings, such as airports and shopping centres and in enterprise installations typically serving at least 100 users. In public environments services need to be provided by multiple operators, and the DAS allows many operators to share one system and deliver multiple services and frequency bands, such as 2G and 3G. In this way they avoid excessive visual disturbance and cabling.

However, distributed antennas require specialised design and commissioning to work effectively, with considerable RF expertise involved. Passive systems have little flexibility and frequently involve the use of large-diameter coaxial cables to deliver sufficiently low loss in large buildings, which may not be well accepted by building managers. Active systems are easier to install and capable of extended reach to long distances and even multiple buildings, but require consideration of the noise budget and power budget of the system. They require power locally at the antenna units and at various intermediate units, careful management and maintenance and typically have a somewhat limited bandwidth or capability for supporting high capacities. See Chapter 13 of (14) and (24) for further information on the detailed engineering aspects of the DAS.

Of most relevance when considering femtocells, the capital cost of the DAS is dominated by the costs of a combination of the professional services involved in expert design and the commissioning and installation of the often very complex cable runs involved. The source which feeds the DAS is typically a full-scale macrocell base station in order to provide sufficient capacity, but with most of the transmit power dissipated in attenuators. Its operational cost therefore includes all of the backhaul, power and site rental (in the case of public buildings) costs of a macrocell. So while the DAS has a definite and significant niche in delivering coverage to larger buildings, it has not been successful in addressing the needs of smaller buildings. In particular, there is a substantial need to find solutions which are cost effective for the very large number of small enterprises.

2.3.4 Wireless Local Area Networks

Wireless LANs, based mostly on the IEEE 802.11 standards (25) and the interoperability specifications of the Wi-Fi Alliance (26), have become very successful, and are unquestionably the most widely deployed example of small cells internationally today.

A number of factors have contributed to this success:

- Demand for data: the need to move large quantities of data between devices, particularly between laptop computers and the Internet, has created a clear demand for wireless data.
- Support for devices: the incorporation of Wi-Fi enabled chipsets into a large proportion of laptops has both encouraged user take-up and enabled a reduction in the costs of the associated chipsets.
- Regulation: regulations have enabled the use of multiple spectrum bands (on an unlicensed or licence-exempt basis) and permitted these bands to be used for revenue-generating services, paving the way for public Wi-Fi services provided by operators.
- Support for home users: although wireless LAN standards were originally created to support the local area networking needs of enterprises, they did not truly take off until home users adopted them. The associated economies of scale and the user familiarity then enabled them to re-emerge for their intended purpose in the enterprise.

Interestingly, most of these factors are also characteristics of femtocells. In the case of device support, the number of mobile devices which can access femtocells far exceeds the number of Wi-Fi embedded devices, though there is a clear distinction in the dominant device type, with mobile devices typically being small and personal handheld devices rather than laptops. This distinction is reducing over time, with some phones supporting Wi-Fi (albeit a very small proportion of the total) and an increasing number of mobile computing devices having embedded mobile broadband capability. With the emergence of dual-mode devices, both at the user end and in the gateway, it is likely that the two technologies will coexist for a long time to come. For the deployment of mobile services by mobile operators, femtocells are the more natural choice as explained in Section 1.6.

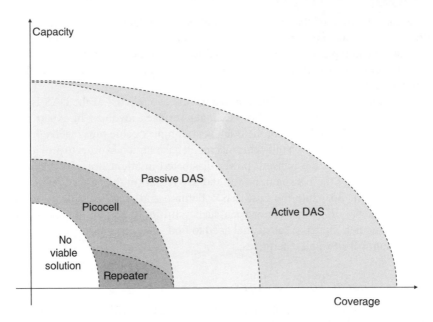

Figure 2.7   Options for dedicated in-building coverage prior to the advent of femtocells

2.4 The Small-Cell Landscape

Figure 2.7 indicates how the choice amongst small-cell solutions by operators has appeared to date. Distributed antenna solutions are the approach of choice for most large-scale buildings, with active systems used in the largest buildings or where multiple buildings are to be served from a single source. Picocells have been used in mid-size office buildings where coverage from macrocells is not available or there are particular service reasons to adopt a dedicated small-cell approach. Repeaters have been used in locations where simple coverage is the issue, but tend not to be favoured by operators in large numbers due to the difficulty in managing them and the negative impact on network capacity. There is a clear gap in solutions at the smaller end of both the capacity and coverage scales, corresponding to home and smaller workplaces. Such environments constitute a large proportion of the market and account between them for potentially around half of an operator's traffic.

Figure 2.8 illustrates how the advent of femtocells and other technology developments expands the range of options for operators and fills the gaps amongst previous solutions. The small coverage, small capacity needs of homes and small offices are certainly addressed. The opportunity for repeaters is likely to diminish as operators find that the controllability and capacity advantages of femtocells address concerns with repeaters – and at much lower hardware and installation cost. At larger scales of capacity and coverage, cooperative networks of femtocells and/or picocells, evolved to include femtocell principles, will impact on the traditional space of picocells and the smaller distributed antenna systems. In public buildings, the DAS is likely to continue to play an important role, due particularly to its support for multiple operators and multiple technologies. At the top end of the scale in capacity and coverage, distributed base stations, evolved from base station systems, with digitally fed remote radio heads, may impact on the DAS market. These utilise developments from both

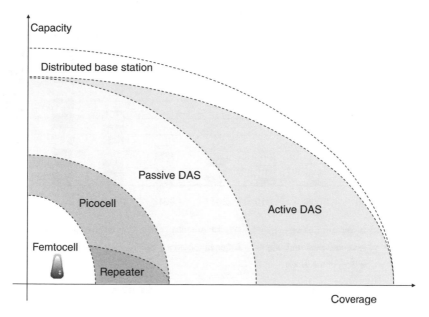

Figure 2.8   An expanded set of options arising from femtocells and other developments

base station hotel and femtocell concepts, but it is uncertain at this stage whether these will achieve the necessary economies of scale for widespread use.

In interpreting Figures 2.7 and 2.8, it is important to recall two points:

1. Although the smaller capacity and coverage environments occupy a small part of these charts, they constitute an enormous untapped market. Small- and medium-size businesses, being hugely more numerous, contribute at least as much to many economies as larger businesses.
2. Although there may be some overlap between the top end of the femtocell market and the bottom end of the DAS market, these are not really competing technologies, since they play a very different role and are addressing very different locations in the main.

2.5 Emergence of the Femtocell – Critical Success Factors

The absence of solutions for coverage and capacity in homes and smaller enterprises has created a substantial need for femtocells. There have been several small-scale projects within several major cellular infrastructure vendors for very small base stations, dating back to at least the mid-1990s, yet these have failed to gain market traction. So why have femtocells only now become a commercial reality?

It appears that several factors have together established both the need and the potential for femtocells at roughly the same time. In the absence of these, femtocells were simply not commercially viable previously. These factors can be summarised as:

• Mobile data adoption – and the associated revenue growth.
• The adoption of fixed broadband.

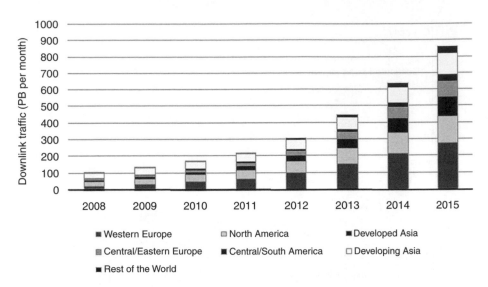

Figure 2.9 Forecast growth in downlink traffic on mobile networks, from (2). *Reproduced by permission of Analysys Mason Ltd*

- The large number of mobile users.
- The emergence of compelling mobile Internet applications.
- The transfer of traffic from fixed networks to mobile networks.
- The availability of compelling user devices supporting data applications.
- The availability of processing power at sufficiently low cost.

Each of these factors will now be examined in turn.

2.5.1 Mobile Data Adoption and Revenue Growth

As has been noted before, 2007 saw a dramatic acceleration in the adoption of mobile broadband services and this trend is expected to continue to grow substantially as indicated by many forecasts such as that shown in Figure 2.9. Helpfully, this growth in data is also accompanied by substantial (though not proportional) growths in data revenue, helping to offset the decline in voice revenues in many markets and to enable continued overall industry growth.

2.5.2 Broadband Adoption

In order to deliver smaller cells of any type, access to cost-effective backhaul solutions is essential. This is becoming increasingly challenging as the demand for data volumes rises. The high cost of traditional leased-line backhaul is causing many mobile operators to switch to microwave links and Ethernet-based solutions for macrocells, but these do not scale downwards in cost sufficiently for home or small-business markets. However, the growth of adoption of

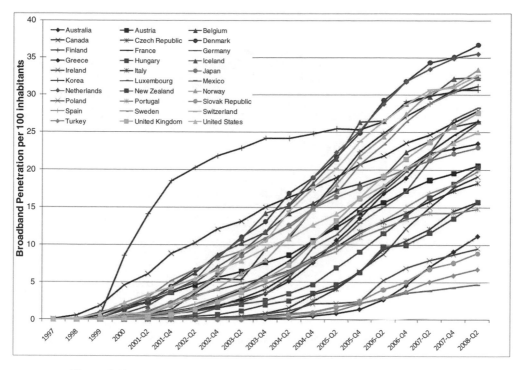

Figure 2.10   Growth of broadband penetration, based on OECD statistics (27)

home broadband, whether based on DSL, cable or fibre solutions, has been impressive in recent years, as shown in Figure 2.10, creating the opportunity for cells that directly reuse a resource, which is already being adopted and paid for by the end customer, delivering additional value for those customers from their existing spending and increasing the efficiency of service delivery for mobile operators.

2.5.3   Connecting Four Billion Users – And Counting

The number of mobile subscribers worldwide passed three billion in April 2008, was at over 3.8 billion at the time of writing (January 2009), and is likely to have passed four billion by the time this book is published, thereby accounting for nearly 60% of the world's population. See Figures 2.11 and 2.12 for a breakdown of these users by technology and geography. This is expected to rise to 5.63 billion by 2013, accounting for 80% penetration by population (28). Femtocells are being designed to support virtually all of these users without modifying their existing devices, thereby enabling service to an unprecedented user base for any technology. Perhaps not all of these devices will be supported to the same degree (18% of the current total is on 3G) and there are challenges in market introduction, including price points and the availability of broadband, but the need to provide good service to these users when they are at home or in the workplace is pressing and represents an attractive commercial opportunity on a large scale.

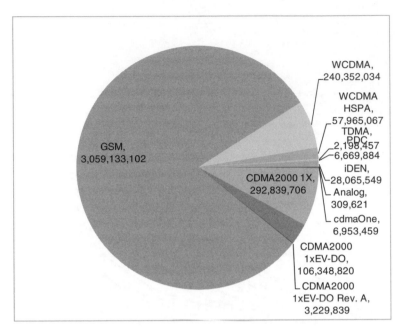

Figure 2.11   Worldwide mobile subscribers by technology, based on statistics from GSMA, September 2008 (29)

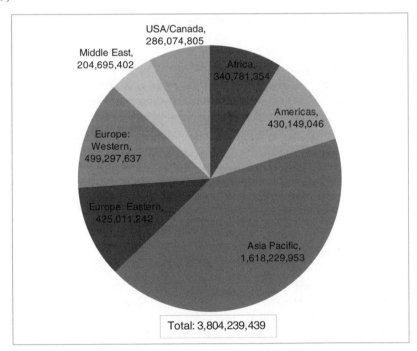

Figure 2.12   Worldwide mobile subscribers by region, based on statistics from GSMA, September 2008 (29)

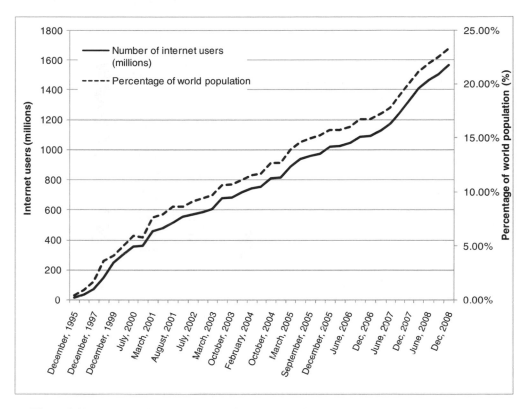

Figure 2.13   Growth of Internet users and proportion of population, based on statistics from (3)

2.5.4 Internet Applications

Use of the Internet in general terms is continuing to grow strongly. The growth in the number of users is shown in Figure 2.13, while the historical and forecast growth in the overall IP traffic is shown in Figure 2.14. Increasingly users want to access the Internet while on the move via their mobile devices and mobile versions of major web sites and applications for Internet search, video, social networking and gaming are in increasing demand. For example, in June 2008, 20.8 million US mobile subscribers and 4.5 million European mobile phone subscribers accessed search during the month, an increase of 68% and 38% from June 2007, respectively. The UK had the highest penetration of mobile subscribers using mobile search at 9.5%, followed closely by the United States at 9.2% (30). Similarly, the use of mobile Internet applications is growing, enabled by smartphones and flat-rate data plans – see Figure 2.15 for example.

2.5.5 Fixed–Mobile Substitution

Although the number of households with access to broadband is rising rapidly as shown in Figure 2.10, the number of computers per head of population is reaching a plateau in many markets as shown in Figure 2.16. This coincides with the continuing use of mobile in preference to fixed-line services for voice calls, such as the example for the UK in Figure 2.17.

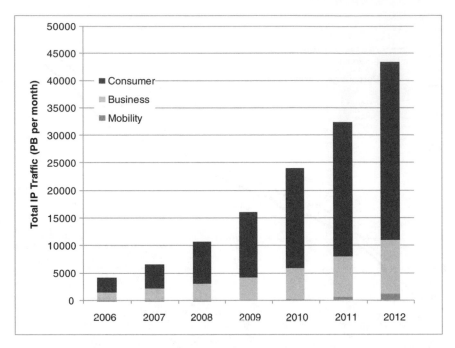

Figure 2.14   Growth in global IP traffic, based on data from (4)

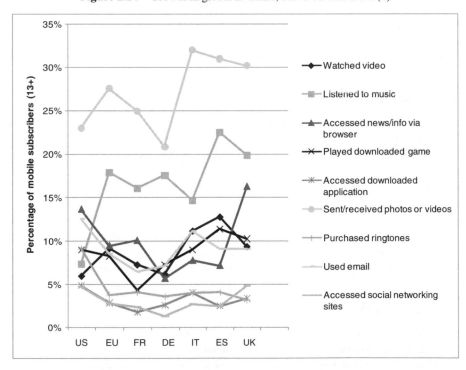

Figure 2.15   Diverse use of mobile Internet applications, based on data from (31)

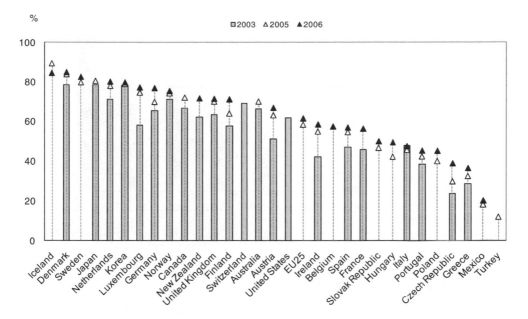

Figure 2.16   Percentage of households with access to a home computer, based on OECD statistics (27)

Thus, there is a clear opportunity for continuing growth in access for individuals to be via personal, mobile devices rather than via computers or tethered telephones. It will come as some relief to fixed-line operators that femtocells make clear that the fixed line in the home should be retained, even for access to mobile services (see Section 11.3 for more on the impact of femtocells on fixed-line operators).

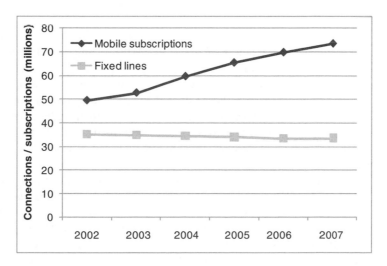

Figure 2.17   Comparison of mobile and fixed-line subscriptions in the UK, based on Ofcom data (32)

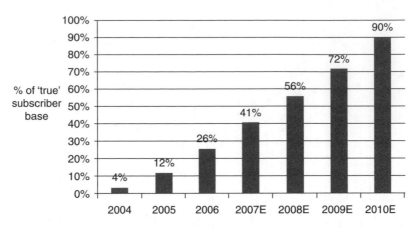

Figure 2.18   Adoption of 3G handsets in Europe, North America and Japan, based on data from (6)

2.5.6 User Device Availability

We have already noted the large – and still increasing – number of mobile users around the world. Of these, currently most have access to 2G services, yet the majority of femtocell activity is in 3G. This is enabled by the rapid growth of 3G handset adoption, as illustrated in Figure 2.18. Thus, even customers who only make use of voice services, can increasingly be enabled by femtocells, and it is clear that over the longer term it is safe to assume that 3G handsets (and beyond) are the best target for new services.

As importantly, these handsets and other user devices enable and encourage use of rich data services by the adoption of suitable user interfaces, including touch screens, high-resolution colour screens, high-quality music capabilities and large quantities of memory.

2.5.7 Processing Power and Cost

A third-generation base station requires of order 300 MIPS – millions of instructions per second – per channel (33). This is dominated by the signal processing needs of the physical layer processing, especially for the receiver, where functions such as correlators and Rake receivers have to operate at very high speeds. Although the processing required for a femtocell is reduced compared to a macrocell base station – primarily due to the reduced capacity and – given the right standards – a reduced need to equalise the multipath channel – the cost involved hitherto made such deployments unthinkable within price points that would be acceptable for home installation.

In around 2000, a macrocell three-sector base station had a bill-of-materials' cost of order $150,000, of which at least 25% was baseband silicon. Moore's law, and variants of it, act to make a substantial improvement: Moore's law in its basic form observes that the number of transistors which can be inexpensively placed on an integrated circuit doubles approximately every two years (34), (16). Other formulations of Moore's law suggest that the price for a given processing power halves every 10 months (35). New processing architectures, efficient algorithms, appropriate standards and large volumes are all required to realise sufficient price reductions for domestic use, but these are all now within reach in a way which was simply impossible even a very few years ago.

2.6 Conclusions

In this chapter, we have seen that the concept of small cells enabling increased usage of spectrum is nothing new, but has been a feature of wireless communications since it began. The motivations for small cells include the need to serve increasing numbers of users, increasing data rates and increasing data volumes. Small cells succeed in meeting these needs by a combination of denser frequency reuse, better protection against interference and the use of lower cost backhaul. Small cells include a wide range of existing approaches, including microcells, picocells and distributed antenna systems, all of which serve particular application requirements. Yet there has been a significant gap in the ability of existing approaches to serve the smaller locations, including homes and small offices, and only now has the technology and commercial motivation existed to address this problem on a wide scale. The emergence of the femtocell arises from the alignment of several factors simultaneously – a syzygy[2] – at the present time, creating the chance for a solution to a compelling and urgent problem.

---

[2] A *syzygy* (36) is a conjunction of heavenly bodies - i.e. the stars are now aligned for the emergence of femtocells.

3

Market Issues for Femtocells

Stuart Carlaw

This chapter aims to give insight into the major market-related aspects that impact upon the uptake of femtocell solutions. Some of the ground touched on in other sections is covered, but here these factors are related specifically to what impact they have upon the market.

3.1 Key Benefits of a Femtocell from Market Perspective

This section includes extracts from (36).[1]

3.1.1 In-Home Coverage

Buildings are one of the most difficult environments in which to propagate radio signals. Building walls absorb RF energy and bounce signals around to produce multipath, one of the key signal killers in any radio environment. There are ways to mitigate the effects of poor wireless coverage, such as the use of leaky feeders or simply raising the power output. Other solutions include distributed antennas, repeaters and picocell base stations. All of these are significantly different from the method employed by femtocell base stations, and the difference in approaches holds the key as to why femtocells are more suited to certain situations.

The overriding difference between femtocells and other solutions is the location of the connection of the femtocell to the fixed assets for backhaul and power. All of the other methods still rely on the cellular carrier's assets to backhaul the signal through either point-to-point radio or leased lines. Another major difference is power. Femtocells will be placed in small offices and houses and the cost of powering these devices will therefore be absorbed by the consumer. Effectively, the operator will be getting the benefits without having to pay for the costs.

Femtocells satisfy the need to provide coverage in the household in the same way that a low-end repeater does. However, there are two major differences between the approaches. The repeater repeats only the signal as received from the network, so a unit in a 2G coverage

---

[1] Reproduced by permission of ABI Research Ltd.

Femtocells: Opportunities and Challenges for Business and Technology   Simon R. Saunders, Stuart Carlaw, Andrea Giustina,
Ravi Raj Bhat, V. Srinivasa Rao and Rasa Siegberg   © 2009 John Wiley & Sons, Ltd

area will repeat only a 2G signal. This holds little value for an operator looking to boost data ARPU. The 3G femtocell actually creates another 3G mini-cell in its location; operators can take advantage of the added revenue from data applications that can be supported in that location. This is especially important in rural areas where 3G coverage is sparse or nonexistent.

Previous research by ABI Research has indicated that 28% of consumers were interested in femtocells from the prospect of indoor coverage alone. The attraction of improved coverage in the home cannot be overestimated. It is also important to note that the evolution to HSPA+ and LTE, as well as the move to higher frequency bands, means that in-building coverage will continue to be a major issue.

3.1.2 Macro Network Capacity Gain

In urban areas, the average load on a base station can be extremely high, with little available capacity. As data use per head increases, the average load will also increase to the point where network operators will be faced with the difficult question of how to negate cell shrinkage. One method is to simply add more network assets in the form of cards, elements or base stations themselves. This is an expensive undertaking that cuts into operating costs.

Femtocells offer network carriers the opportunity to offload a lot of the traffic from their network assets to subscriber home-based cells that are backhauled through the IP core. This is a special consideration given the increasing amount of time consumers use cellular handsets in the home. The operating expense savings from not having to cater for voice traffic alone provides enough motivation to build femtocells into the business plan. When this is combined with the distinct possibility that the use of femtocells will delay the need to add network assets due to cell shrinkage and traffic load, then the motivation for operators to look seriously at femtocells is high.

Operators expend a considerable amount of capital in satisfying backhaul requirements. Currently, they have only two main options for backhaul. The first is to lease E1 or T1 lines from the incumbent fixed-network supplier. This is a relatively expensive exercise as it does not take into account fluctuations in capacity and use and is an ongoing cost that will have little chance of decreasing. It is also important to consider that operators are often forced to purchase leased-line capacity from competitors that hold both fixed and mobile assets.

The second major option for backhaul is the use of microwave radio. This significantly reduces the long-term operating expenses associated with leased lines, but also has some other significant drawbacks. The solution involves the up-front investment of a major amount of capital to buy the radio equipment and to secure the license. There is also the ongoing cost of running and maintaining the equipment (power costs and breakage).

These points illustrate that operators have a lot to gain by offloading as much traffic onto the IP core as possible. In the case of the femtocell, it is important to note that it is not only a matter of offloading the backhaul traffic to the IP network; the model is partially funded by the consumer and other services such as broadband and IPTV have the potential to be piggybacked on top of the service.

3.1.3 Termination Fees

Carriers will benefit from added revenue from incoming calls that are terminated on their own network assets. Essentially, what would have been termination fees paid to fixed-network incumbents such as BT for a fixed-line call or a mobile call terminated at a fixed phone would now be paid to the mobile carrier.

3.1.4 Simplistic Handset Approach

One of the key differentiators of the femtocell solution in comparison to the VoWi-Fi (also called dual-mode handset elsewhere in this book) solutions is that the handset side of the femtocell offering is simplistic. Handsets do not need any client software on board to support FMC. This is a major advantage for the femtocell offering as one of the major bottlenecks restricting the uptake of VoWi-Fi solutions to date has been a restricted portfolio of available devices. In theory, carriers will have the complete portfolio of any manufacturer from which to choose, unlike VoWi-Fi, where carriers are generally restricted to three or four devices, although this may increase over time.

One other aspect of the handset equation is related to battery life. Continually searching for macro base stations reduces battery life and, conversely, VoWi-Fi devices suffer from low battery life due to the inefficiency of Wi-Fi for the transmission of voice. Femtocells offer the opportunity to improve both of these situations by refining signal quality and also removing the need for Wi-Fi radios.

3.1.5 Home Footprint and the Quadruple Play

Probably the most pervasive argument for wireless operators to adopt femtocell-based services is enabling a competitive triple-play or quadruple-play solution including mobile service within bundles of other telecoms, broadband and entertainment services. When analysing this motivation, the core aspects of the market conditions and the major factors for success must be taken into account.

The fundamental characteristics of the market that have the major effects are competition and saturation. Increased competition in the communications and entertainment broadcasting markets means that prices are inevitably being driven down, innovation is high, and everyone is striving for differentiation. All these come at a cost, whether it is the expense of a subscriber addition, the decreasing margins due to price decline, or the need to expend more on R&D to stay ahead of the curve technologically.

These factors are further exacerbated by the saturated subscriber base. Competition in an expanding market is always easier than in one that has reached its limits. Operators must win subscribers from competitors, and this means that all of the metrics listed above – price degradation, innovation and differentiation – are magnified.

These conditions lead to a salient set of facts. Operators must maximise their share of consumer spending on communications and entertainment; they must be as efficient as possible in the delivery and support of services; they must minimise churn; and they must continually innovate and differentiate.

The home is one of the few environments where the fight over who will dominate in terms of multimedia services, delivery and distribution is still to be resolved. The competition in this market exists at all levels, from equipment vendors targeting the home to service providers looking to capture more consumer revenue. One of the fundamental problems faced by the cellular carriers is that they lack a physical footprint in the home. This puts them at a significant disadvantage to competitors when supporting a comprehensive FMC and quadruple play. Femtocells provide that much-needed footprint.

3.1.6 Maximising Returns on Spectrum Investment

Carriers globally have paid billions of dollars for spectrum licences, especially 3G, and have yet to see any returns on those investments. Moreover, high handset prices for 3G have resulted in migration of subscribers from 2G to 3G which are well below initial expectations. The fact is that 3G not only provides higher bandwidths for services but also is far more economical in terms of the number of users that can be accommodated on each base station or carrier. Cellular operators can economise on their network investments as they use the spare capacity that is currently sitting on expensive 3G networks.

A more aggressive migration of subscribers from 2G to 3G networks will allow carriers in European and some Asian markets to refarm some of their 900 MHz spectrum from 2G- to 3G-based services. The simple move to a lower frequency will allow for greater cell sizes, fewer cell sites, improved indoor coverage and a general improvement in the economic position for 3G in terms of ROI even in consideration of the large licence fees.

The final point regarding the maximisation of spectrum investment is that, due to femtocells using cellular frequencies, only carriers with licences or access to licences through contractual agreements, will be able to offer the solution. This offers carriers a gilt-edged chance to provide a technological footprint that is significantly differentiated from other competing companies in the fixed, cable and satellite broadcaster spheres. This is important given that the market for consumer spending on telecoms and entertainment is being eyed ever more competitively by a growing band of companies.

3.1.7 Churn Reduction – The Sticky Bundle

One of the major characteristics of any saturated market is that churn becomes a significant issue as subscribers learn of enticing deals offered by competitive carriers in the same market. More importantly, the cost of churn is high and has a major effect on the bottom line. Many carriers in developed markets have focused on churn reduction as one of their major goals.

Femtocells offer carriers a significant vehicle for churn reduction. When carriers position the solution with multiple users all on the same femtocell, it will reduce the tendency for those users to feel the need to move to another carrier as it will mean having to replace the box and all of the handsets. This could be costly for the consumer and may mean the loss of a favoured aspect of the service. Also, the addition of family groups onto a single femtocell will discourage single members of that group from splintering off and joining other carriers. Typical teenagers will see femtocells as a favourable option, since their phone services will be funded by their parents. Carriers are likely to be creative with family pricing plans in order to seed the market for this type of situation.

Churn promotion is another issue. Carriers that are first to market with femtocell solutions are likely to attract subscribers from other carriers. Once in place, if the service is executed properly, customers will be unlikely to go to other carriers until those carriers can offer comparable solutions. Once a carrier in a certain market offers a service and begins to win share, other carriers will have to move quickly in order to remain competitive. This is likely to foster a positive reinforcement cycle that will speed uptake.

3.1.8 Positive Impact on Subsidisation Trends

As mentioned in previous sections, carriers are continually looking to bring down the cost of subsidisation since it often is not outweighed by the resulting revenues generated from users. This is especially true in the case of 3G networks where handset prices are higher, subsidisation is most costly and user ARPU has previously been only marginally better than for 2G subscribers. Carriers have tried to minimise the cost of subsidisation through different methods including providing low-cost 3G handsets made by ODM partners and also by reducing the subsidies paid on expensive handsets.

Femtocells can enable carriers to further reduce the subsidisation on handsets. They can offer fully subsidised femtocells and handsets but recoup more than the value of the total subsidisation through monthly subscription fees for the femtocell service alone. A $25 per month subscription fee over an 18-month period would generate $450, much more than a cost-optimised femtocell and four cost-optimised handsets.

3.1.9 Value-Added Services

Probably the most interesting recent aspect in the femtocell market is value-added services. Most femtocell vendors want to incorporate as much intelligence in the access points as possible.

This high level of intelligence in the box allows essentially limitless new services to be layered into the package. The market is still in the early stages and the majority of business models are built around the idea of voice and data and are yet to consider what other applications can be delivered to the user. It is likely that value-added services will take some time to emerge, and more important, the exact nature of the services that will be most prevalent is still unclear.

Some potential applications have already been identified and are likely to be in the first wave of value-added service releases. They include the following and the other examples provided in Chapter 11:

- remote home control, security and automation;
- incoming call routing;
- remote home control and automation;
- user location and inter-house calling;
- home security monitoring and control;
- media distribution and the ability to shift content to mobile devices from a femtocell.

All of these applications are embryonic in terms of the practical reality of what they will finally encompass, but one thing is for certain – these services will reinforce the positive trends promoted by femtocells in terms of ARPU as well as subscriber retention.

3.1.10 Changing User Behaviour

Also important is the ability to change consumer behaviour, especially in view of 3G data services. Until now, data ARPU has been significantly lower than expected and this is contributing to a worsening economic picture for carriers in some developed markets. Putting femtocells into homes allow carriers to price data services attractively in order to stimulate users' exposure to newer data-centric services.

It is hoped that this exposure will result in consumers opting to use advanced services more often. It is also hoped that this change will spill over to the macro network where premium or higher prices will be charged with a significant compound upside on data ARPU.

3.1.11 Reducing Energy Consumption

The top three costs that wireless carriers face are site rental, backhaul expenses and energy consumption. Energy consumption is especially prominent since it is subject to variation based on global energy markets with margins of risk that are beyond the carriers' control.

Carriers are in somewhat of a 'Catch-22 situation' in terms of energy consumption. They need to stimulate mobile data revenues and the most likely way of doing that is attractive pricing plus compelling services. In some cases, this might entail simpler 'all-you-can–eat' pricing bundles. This is likely to bring an upsurge in data use with a negative impact on the network in the form of urgently needed capacity or technology upgrades to take advantage of the heightened traffic levels. Installing new base stations or upgrading macro sites to HSDPA or HSPA+ or even LTE is expensive and seriously challenges the carrier's financial well-being if it is not accompanied by an associated growth in service revenue.

Femtocells are an attractive option in that they off-load traffic from the network and also the cost of energy from the carrier to the consumer, reducing the total required for a given traffic level. Femtocells add massive amounts of capacity in a model that allows the carrier to let the user fund both the capacity upgrade and the energy costs associated with adding more elements to the network.

3.2 Key Primers

3.2.1 Broadband Penetration

Broadband penetration is probably the most important primary driver that is relevant to the femtocell market. The broadband connection forms the primary backhaul method between the access point (AP) and the core network of the carrier via the controller or gateway if present. Therefore, the overall penetration of broadband into households and places of work can be seen as the ultimate limiting factor on market size.

According to ABI Research (37), broadband household penetration reached 190 million households in 2006 and is predicted to rise to over 350 million households in 2013. This clearly indicates that there is a significant footprint upon which femtocell solutions can be rolled out.

At this point, it is also important to note a couple of major factors regarding broadband penetration and how this relates to femtocells. Firstly, regional distribution of broadband has a direct impact upon deployments. It is important to remember the majority of these DSL subscriptions are actually in developed markets and this may limit the penetration of femtocells in developing markets.

The second major concern regarding broadband penetration is the quality of the connection. In markets such as South Korea and Japan where connection speeds up to 100 Mbps are available this is not an issue, but some parts of the developed markets such as the UK see much lower speeds. First of all, the peak rate for DSL can be in the region of 10 Mbps on average and then contention ratios of 20:1 are common, as well as are speeds of less than 1 Mbps on the uplink due to distance from the exchange – see Chapter 5 for more details. This can cause a situation whereby the major limiting factor in femtocell service delivery is not the data rate of the air interface but the uplink capacity of the consumer broadband connection.

One last note regarding broadband connection is that DSL is not the only option. The world sees significant deployments of other technologies such as cable and fibre that also can provide a backhaul solution for femtocell solutions. These solutions tend not have any of the service limitations of DSL but are far less common in the market. ABI Research (38) predicts that by 2013 there will be 248 million houses serviced by fibre to the home – up from 46 million in 2006. Cable connections provide a more significant opportunity for femtocell penetration with ABI Research (39) forecasting 432 million subscriptions in 2013, up from 154 million in 2007.

3.2.2 Saturation

Saturation is a very complex concept that has far ranging and profound effects on all markets. Saturation refers to the condition where mobile subscriptions reach, and potentially surpass, the population of a specific market. Although this, on some grounds, seems a little of an excessive situation it is very much a reality in many developed markets. ABI Research predicts that by 2013 many of the global regions will hit and surpass 100% penetration rates for mobile services. Table 3.1 illustrates some interesting penetration rates for selected global regions.

The penetration rates exhibited in Table 3.1 clearly illustrate that saturation is occurring. Before discussing the impacts of saturation it is also important to consider that saturation is a relative term. Saturation in developed markets is rightly thought of as in relation to the total population, however, this is not so valid for the emerging markets. In emerging markets the natural target for femtocell solutions is the higher spending subscribers. These tend to only account for a limited number of the total base. For example, it is widely estimated that the Chinese market can be split into two very different markets – 35% at the top which are relatively high-spending urban subscribers and the rest which represent a lower value but far larger group of subscribers. In this situation saturation must be viewed in relation to the saturation of service subscriptions into the target markets.

Moving on to the effects of saturation: these are many and profound. Carriers used to measure success in terms of subscriber additions but now success has moved to measures

Table 3.1 Mobile penetration rates

Region	2006 penetration	2013 penetration
Western Europe	108.3 %	139.1%
Eastern Europe	89.7%	135.1%
Asia/PAC	31.0%	64.2%
North America	75.5%	114.4%
South America	53.1%	100.7%
Middle East	47.4%	86.8%
Africa	20.2%	45.6%

Source: ABI Research (40)

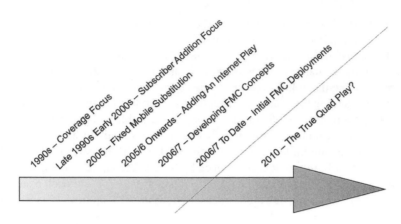

Figure 3.1   Carrier business model focus

of ARPU and retention. Carriers can no longer rely on adding subscribers to support their business models.

One last aspect regarding saturation is that subscribers are beginning to own multiple products. Multiple device ownership presents specific pressures on the carrier community where carriers are expected to support multiple devices with lower associated ARPUs with equally high CAPEX costs in terms of subsidization.

In summary, carriers are forced to find CAPEX and OPEX savings and retention methods at every corner. Femtocells hold the potential to offer these.

3.2.3 Evolution in Carrier Business Model

Much related to the previous point of saturation, the carrier business model is evolving too. As alluded to in the previous section, the pressure of saturation has evolved to where carriers in many developed markets are looking to secure revenues from solutions outside of a pure mobile play. Figure 3.1 gives an outline of the timelines and development of the carrier focus on business models.

It is quite clear that carriers in developed markets, have migrated their primary focus for potential femtocell solution deployments from a purely siloed approach – where they look to capture as much consumer wallet in the mobile communications space as possible – to looking at the whole communications and entertainment spend of the consumer. The current trends in the carrier market can clearly be seen in this, where carriers are moving to add services such as the Internet and TV to their solutions package in order to succeed in this notion.

Femtocells hold the opportunity to support this through tying the cell phone into the full quad-play concept at the same time as providing significant efficiencies in terms of the cost of delivering these services. This last point of cost is incredibly important. The move to capturing more wallet and consolidated billing often leads to overall reduced cost of the subscriber and reduced overall revenues for the carrier. It is therefore absolutely imperative that carriers find the most economic solution of providing these convergent services. Femtocells can provide this.

3.2.4 Competition

The concept of evolving business cases to where the carrier is focused upon capturing more wallet share across not only the wireless space but also the complete communications and entertainment spend is an incredibly important driving factor. Up until this section the majority of the discussion has been about how mobile carriers are changing their business models to better address adjacent markets, however, it must also be recognised that carriers in adjacent markets are eagerly eyeing the opportunity to move into the wireless space.

Examples of this would include the Clearwire deal in the United States where cable companies now have a route to mobile services. More traditional examples can be seen in BSkyB (a satellite TV provider) in the UK where they now have a successful MVNO arrangement with Vodafone.

It is natural that, as these major suppliers encroach, competition will not only grow, it will also be multiplied by a layer cake of different companies each bringing different business models and disruptive approaches to the market. Figure 3.2 gives an illustration of how this layer cake of competitive entities stacks up over a period of time.

Carriers in this environment must look to future-proof and differentiated solutions that clearly set them aside from the competition. Carriers must also react quickly to competitive moves from other companies. In such a market, the attractive concept of a femtocell solution

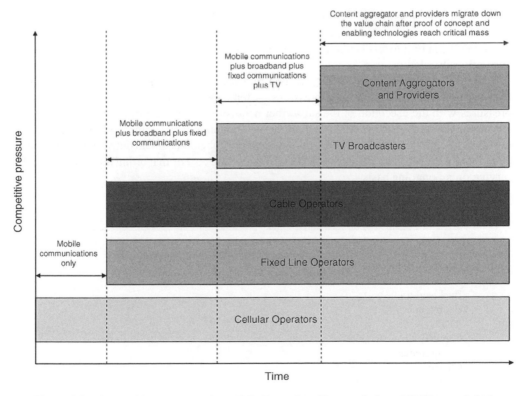

Figure 3.2 Competitive pressures, from (39). *Reproduced by permission of ABI Research Ltd*

could be launched by one carrier and then competitive carriers would have to launch comparable services at their earliest convenience. This is already being seen in France and the United
States where Orange and T-Mobile respectively have launched UMA-based dual-mode Wi-Fi
solutions and other carriers are now launching or are on the brink of launching competitive
femtocell solutions.

3.2.5 Technical Feasibility

The femtocell concept was first considered back in the late 1990s and early 2000s but was
deemed as incapable of supporting a viable business model due to the lack of demand and the
inability to make a product that worked well at the same time as being affordable for enough
consumers to be able to purchase the product or be cheap enough for carriers to subsidise as
discussed in Section 3.4. The question must be asked: what has changed?

The simple answer to this is that Moore's law has been implemented with great effect in
the silicon markets and has resulted in the processing power per dollar being low enough to
make the solution cost effective but also have the processing power able to handle the complex
aspects of femtocell operation. This is especially related to the zero-touch self-optimising
nature of the product where it automatically senses its environment and adapts accordingly.

3.2.6 Economics

Economics have a significant effect upon the route carriers choose to take in any market. It is
imperative to put all of the economic aspects together when considering the prospects of how
a femtocell can fit into this landscape. Table 3.2 illustrates the economic pressures carriers are
currently subjected to.

Table 3.2 clearly illustrates that there is a dire need for carriers in developed markets to
find solutions that ease all of these pressures. A femtocell solution can address most of these
pressures with the exception of the subscriber acquisition costs where it actually exacerbates
the issue. The real question when looking at the economic factors surrounding femtocells is
whether the cumulative financial benefits associated with femtocells can outweigh the negatives
associated with subscriber acquisition costs and also the cost of marketing and supporting the
service on an ongoing basis.

Table 3.2 Economic pressures on carriers

Economic factor	Direction of pressure (i.e. costs more or less)
CAPEX requirements for new networks	↑
Backhaul traffic costs	↑
Energy costs	↑
ARPU	↓
Capacity-driven CAPEX requirements	↑
Churn	↑
Subscriber acquisition costs	↑

Table 3.3 Comparison of alternative fixed-mobile convergence solutions

Factor	Femtocells	Dual-mode	Homezone	Picocells	DAS
Network capacity	*	*		*	
AP cost			*		
Handset availability	*		*	*	*
Handset cost	*		*	*	*
Time to market		*	*	*	*
Standardised	*	*		*	*
Promotes new revenues	*			*	

3.2.7 Limitations in Other Services

There is no doubt that there is a general desire for a fixed-mobile solution that can provide a solution that supports fixed-mobile substitution but also provides a launch pad for future services centred around the quadruple-play. In the same sphere as femtocells we can see a number of other solutions. These are Wi-Fi based dual-mode solutions such as UMA, home-zone solutions such as ZuHaus in Germany, picocells and also passive antenna systems such as distributed antenna systems. Table 3.3 outlines a comparison of the relevant solutions.

In order to summarise these competitive shortcomings the bullet points below give the major reasons behind the limitations in adoption of the differing solutions:

- Dual-mode devices – Lack of device availability and new data revenue, and also that data traffic ends up outside the carrier space due to the use of Wi-Fi.
- Homezone – Short term competitive measure that boosts FMS but costs a lot in terms of added backhaul costs with reduced ARPU. Does not add network capacity.
- Picocells – Costly to install and in terms of equipment. Really only applicable to enterprise and municipal environments.
- DAS – Expensive and not suitable to consumer markets.

3.2.8 Carrier and Manufacturer Support

One of the most significant aspects that has driven the femtocell market is the level of support seen most notably from the carrier community and also from the OEM community. This market is best characterised as seeing more carrier pull than most other markets, which tend to be vendor push, with several waves of carriers, including those that have launched early services, those who have launch plans, and those who have seriously evaluated the technology and are considering the right timing for a future launch.

These carriers account for a large percentage of the global subscriber base – for example, carrier members of the Femto Forum accounted for over 1.3 billion subscribers in January 2009. The fact that many of these carriers are not only actively investigating femtocell solutions but are also involved in the Femto Forum and driving the standardisation processes in 3GPP, 3GPP2 and WiMAX Forum clearly gives an impression of the potential scale of the femtocell solution.

Not only does this illustrate the potential in the market but also gives an indication of the pressure that is being placed on the OEM community in order to support carrier demands. A wide range of OEMs and ODMs is currently involved in this market, as are semiconductor vendors and suppliers of gateway, controller and network software elements. See www.femtoforum.org for a list of some of the most prominent players. In summary, it is clear that there is a lot of supplier momentum behind the solution and it is also somewhat of a self-reinforcing circle in that the more support in the solution, the more is automatically attracted. This drives innovation and provides reach to otherwise unattainable markets.

3.2.9 Consumer Demand

The aspect of consumer demand is one that is less easy to gauge. There are no similar solutions in the market that can be used as an example when trying to survey consumer attitude. In order to portray the solution it is only possible to produce a valid response from consumers when the benefit of the femtocell solution is described as one purely centred on better in-home coverage without the other aspects like cost saving etc. Figure 3.3 illustrates the results of a survey conducted by ABI Research in January 2007 to gauge consumer demand (40).

The figure clearly shows that, even when sold on the precept that the solution just boosts in-home coverage, there are a large number of potential subscribers in the 'likely' categories. It is also apparent that there is a large number of respondents in the neutral segment that can easily be swayed by better market messaging or a more developed service offering.

It could be argued that consumers are agnostic in terms of the type of technology they wish to use and really are sold on service bundles and lifestyle benefits. Essentially, the consumer is interested in purchasing an FMC solution, no matter what technology supports the service. As such, the onus of technology choice then falls on the carrier.

3.2.10 Supporting the Data Boom

There has been a pronounced level of growth in the amount of data being consumed by mobile subscribers. This is partly due to more competitive prices in flat tariff and all-you-can-eat bundles, but also to a large explosion in the number of subscribers to mobile broadband solutions using USB dongles in laptops.

This data boom puts significant pressure on the carrier community in terms of network capacity usage and the need to upgrade capacity in order to meet demand. This is obviously a very costly exercise. Femtocells hold the promise of allowing carriers to put capacity where it is required in a very cost-effective way.

3.2.11 Growing Standardisation

One of the oldest criticisms of the femtocell effort was that it was not standardised in any way. The advent of the Femto Forum has provided a very focused forum for discussion, which has provided significant input and recommendation to the various standards bodies. This has allowed some consensus to emerge on important aspects like the network interface. This consensus allows for the opportunity of multi-vendor environments but also for an economy of scale to be generated whereby suppliers and carriers can really monopolise on the price declines that are associated with significant volumes of product shipments. See Chapter 8 for more details of the progress achieved.

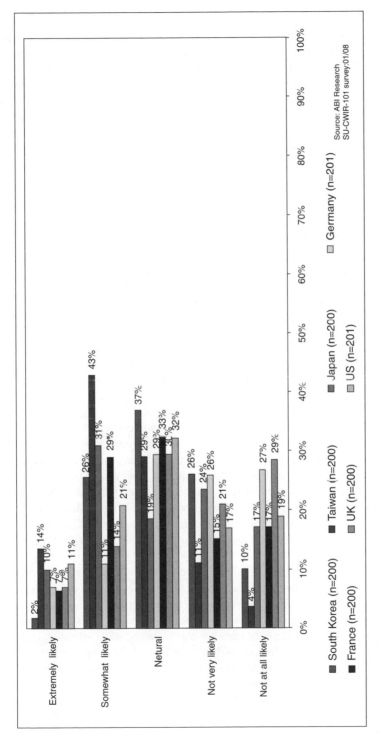

Figure 3.3 Consumer demand for femtocells, from (40). *Reproduced by permission of ABI Research Ltd*

3.2.12 Air Interface Technology Evolution

The cellular communications markets are currently in a period of flux where the carrier community is moving from 2G through 3G to 4G technologies. This all requires a significant amount of CAPEX investment, which can be mitigated somewhat by the use of femtocells, but more importantly, femtocells will be part of the network plan of the future for these 4G technologies. The newer technologies are likely to operate primarily in higher spectrum bands with lower propagation footprints. This is likely to result in poor indoor coverage. The real advantage of these 4G technologies is in higher data rates for mobile data solutions and when this is combined with the fact that the majority of mobile data is consumed indoors where there is poor radio propagation, it becomes imperative that a cost-effective in-building solution is provided.

3.3 Key Market Challenges

3.3.1 Cost Pressure

Probably the primary market challenge that the femtocell offering is facing is cost related. There is a need for femtocell costs to be reduced to appropriate points as quickly as possible in order to enable the price points to be low enough so that the economic situation allows femtocells to be affordable for lower tiers of carriers' subscriber bases.

When looking at the price points for femtocells, it is interesting to note that significant erosion in the target price to operators occurred even before any WCDMA or GSM product has been shipped. Two years ago, $200 price points were mentioned, a year later the target dropped to $120 and now the consensus is around $100 being an achievable target in the short to medium term, assuming sufficiently high volumes. Even after this aggressive decline in demands, there are still those that are arguing for yet lower costs. Some voices in the carrier community are calling for prices of around $75 in the near future or even as low as $40.

This type of price pressure is a natural occurrence in any market as carriers naturally try to improve their financial position to the optimum, however, it raises some significant problems for the femtocell market. Firstly, it acts to make vendors seriously compromise on feature sets. Rather than an integrated gateway product with Wi-Fi, stand-alone products are initially more likely. This is something that is counter to the notion of controlling the whole signal chain in a consumer home and not burdening the consumer with too many in-home boxes.

The second major impact of this significant price pressure is that it reduces diversity from the semiconductor market. Semiconductor vendors see the biggest pressure in the price battle, which tends to involve pressure rolling downhill to the bottom of the chain. At the same time they also see the need to invest huge sums into development costs to provide a working femtocell solution in a market where volumes are unproven and standardisation is at a relatively early stage. This all means that the market is dominated by smaller chip vendors and the largest companies are sitting out the early stages whilst waiting for volume before entering the market through acquisition or commoditisation.

One last point regarding cost pressure is that it is forcing some of the smaller vendors into forward pricing of products. This could be a dangerous situation whereby if there are any stalls in the market they will be very exposed to financial strife.

3.3.2 Intellectual Property Rights

Intellectual property rights could have a dramatic impact on this market. Companies with large 3G patent portfolios will look to leverage their patent holdings, and there are questions concerning what licence model will be used, what cumulative rate will be used and what effect this will have.

When looking at the licensing model, recent precedent has indicated a downstream model (whereby the royalty rates will be levied at the furthest portion down the supplier chain as possible, i.e. at the point when the device is shipped to the end customer) will be most likely but it must be recognised that the overall cost of the femtocell device will be much more than just the radio components. This is especially so when considering that femtocell functionality will eventually be integrated into set-top boxes and broadband gateways. This seems to suggest that a model that is similar to the cellular modem market is likely, where royalties are levied on modem makers only for the radio portion rather than the whole cost of the product.

When looking at this model it is important to remember that the companies, that the emphasis of royalty payments will be placed upon, are the smaller start-up companies which specialise in femtocells. The companies have significant femtocell-specific IP which could be traded in a bilateral negotiation, however, the other larger and more established parties may have little interest in WCDMA products and will not cross-license to reduce fees. This will put increasing price pressure on these smaller vendors at the vanguard of the market and could stall the market.

The final question is regarding IPR concerns when royalty claims are actually initiated. These are likely to occur once volumes have reached 2–6 million per annum. At this time there will be enough incentive to license from companies shipping in this market (which could be used as a negotiation tool) and by this time a significant deconstruction of the femtocells shipping will have been done so that sound claims for prior art can be made.

When all of these are added together, the worst possible outcome would be that stacked IP is so large that it inflates prices to the point where the smaller vendors are priced out of the market due to the lack of cross-licensable portfolio and the diversity and leading innovation slips out of the market as the baton is taken up by the larger vendors who already have established IP arrangements. This could be seen as one of the primers for the larger vendors purchasing some of the smaller companies.

The femtocell industry can mitigate these risks by coordination within the femtocell community with a view to possibly cross-licensing IPR within the Femto Forum members to provide a block of patents that can be leveraged against potential royalty claimants. It is also evident that there has been significant intellectual property developed by those involved in the Femto domain and this does have significant value in negotiations. As discussed above, a model more akin to the embedded cellular modem market could be advocated. This model also sees the levying of rates one stage further up the value chain in the ODM area.

3.3.3 Technology Issues

There is a multitude of technology issues that could potentially cause problems for a femtocell solution. These include:

- in-band interference or separate band;

- security of communications in the IP domain;
- hand in and hand out;
- remote product configuration;
- zero-touch installation;
- quality of service over consumer grade broadband connections.

The general consensus is that all of these technical issues are solved or are solvable and are well covered in other chapters of this book. The idea that technical solutions can be found to all of these problems, given sufficient financial motivation to do so, holds true in this case.

3.3.4 Establishing a 'Sellable' Proposition

The issue of femtocell costs discussed earlier is one important aspect of establishing a viable overall business model. As more carriers look to roll out femtocells, this is a profound issue as marketers look to find ways to convert the benefits of a femtocell solution into something that is highly saleable to the consumer base. Questions remain between whether it should be an FMS-focused proposition or whether consumers are ready to accept a more complex offering that conveys the messages of improved data services.

This situation must be resolved at the earliest time as it is an essential part of the economic foundation for femtocell solution. Even if the tangible network efficiency and CAPEX aspects make the femtocell solution economically viable, the lack of a consumer 'sell' message will significantly impact the propensity of carriers to actively market this product. Active marketing will be an essential component of making this solution workable as can be seen in the case of the UMA-based Unik service in France where a strong and direct marketing plan really has paid dividends in terms of service uptake.

3.3.5 Disconnect Between OEMs and Carriers

It could be argued that there is a certain amount of disconnect between the carrier and OEM communities. The major disconnect seems to be related to the pace of innovation that each of these groups of companies is trying to enforce on the market. The carrier community is intently focused on getting the solution right for initial launch. They are focused on getting every possible eventuality catered for in order to go to market with a very robust and, as trouble-free as possible, solution.

This is somewhat different to the OEM community, which is really pushing for the next stage of the market. This next stage is likely to include the move to more advanced services, the integration of the femtocell into other equipment, enterprise applications and also to enhanced 'super-femtocells' for municipal location applications.

There is an evident divide in this approach and some may argue that this results in a dilution of common effort. Others may argue that this is simply a normal part of the push and pull between the different time horizons of vendor and operator communities.

3.3.6 Too Much Reliance on Standards

One potential issue that has recently raised its head is that carriers might be too reliant on standards. Carriers naturally want to come to market with a future-proof product that takes the benefit of a standardised environment. However, the standards process is still being resolved. The carriers with the most immediate business needs will roll out with products that are pre-standard in order to get the best out of the current market conditions. Waiting until the solutions are 100% fully standards bullet-proof may well reduce the opportunity for optimum success when entering the market.

3.3.7 Window of Opportunity

There is no doubt that a perfect storm of conditions is building, that is well-suited to femtocell- and FMC-based solutions. It must also be recognised that there is a perceived slackening of interest in UMA and dual-mode Wi-Fi solutions. It could be argued that, if the roll out of femtocells takes too long it will miss this perfect storm and, carriers will move to competitive solutions. This is only exacerbated by the fact that UMA clients and Wi-Fi are becoming cheaper and more common in mobile devices.

3.3.8 Developing the Ecosystem

The last major challenge relates to the breadth of the ecosystem. The early stages of the market were dominated by WCDMA carriers and OEMs and there was relatively less representation from the CDMA and WiMAX market. However, the situation is rebalancing, and it should be recalled that the first commercial femtocell deployment was for CDMA. More interestingly, there is a perceived lack of application developers present in this environment that can aid in proliferating the solution by making the applications more pervasive to the consumer base but also provide a more substantial potential revenue stream.

3.4 Business Cases for Femtocells

The business case for a femtocell launch is one that differs markedly from carrier to carrier and depends very much on the specific conditions of the geographical market that the carrier is operating in. In order to give some insight into this complex issue, the following sections outline the most important business case foundations that a carrier must consider when constructing a business model.

3.4.1 Business Case Foundations

The following aspects form the core focus points that a carrier must consider when putting together a business case:

- Cost and subsidisation of the access point – The cost to the carrier of the femtocell was discussed at length earlier in this chapter, however, the aspect of subsidisation is also important to consider. It is unlikely that consumers would be willing to support the whole cost of a $100-plus femtocell in most markets. Carriers will then be forced into considering subsidisation for products. The level of subsidisation currently being considered by carriers, ranges between 50% subsidisation right through to 100%. It must also be remembered that, if carriers can stimulate volume through subsidisation, their costs will reduce over time as volumetrics drive the average costs of femtocell APs down.

- Delayed CAPEX on capacity-driven network investment – Rather than look at upgrading large macrocells with extra transceivers, capacity can be placed in exactly the point where it is required and with some level of financial support from the consumer.
- Improved macro network quality – Taking subscribers off macro networks is a highly beneficial aspect of femtocell solutions, especially WCDMA products. Taking traffic off the network reduces the noise level and increases the quality of service, which can have a very positive impact upon subscriber retention.
- Increased subscriber retention – Femtocell services will be very attractive in attracting family and group members. It is a very sticky subscriber bundle that will be hard for consumers to move away from easily. It is also a possibility that a femtocell may be used as an incentive to encourage subscribers to commit to longer contract periods of 18 to 24 months.
- Churn-in – The first to launch in any market with a pervasive service based on femtocell solutions is likely to receive some positive churn-in from other carriers. Obviously this is likely to be transitory until other carriers take defensive actions. It is also the case that, as families and groups migrate to a femtocell solution, it is almost always at the initiation of the budget holder who drags others into the contract. Carriers could use positive relationships with current high-value customers to bring in other members of their family units.
- Reduced backhaul costs – The most significant amount of subscriber data traffic (as high as 90% according to some estimates) is consumed in an in-building environment. If a femtocell is used to accommodate that traffic, then the backhaul traffic will go through a consumer broadband connection and will unburden the carrier of one of its top OPEX considerations.
- Delayed CAPEX for new network technology roll-outs – The carrier community is in the process of investing in advanced 3.5G and 4G services which are likely to deployed in hot-zones. Data services will be at the core of these solutions and therefore coverage where services are used is required and also that cost of delivery per bit of data needs to be optimised. Femtocells offer carriers an opportunity to do this in a cost-effective and timely way.
- Support for flat-rate data plans – There has been a noticeable shift in carrier business models to supporting flat-rate data plans. These plans generally result in increased traffic as the cost per megabyte of data drops significantly. Unfortunately, as data prices drop the carrier needs to support more traffic at less margin. Femtocells offer the carrier the opportunity to add a more economical method of supporting flat-rate data plans.
- Cross-selling of other services – As previously covered in the chapter, carriers are looking to expand their offering from those restricted to just offering a single communications service. Femtocells give carriers the opportunity, and footprint, to be able to cross-sell other services that they offer.
- Value-added services – As well as the standard services like mobile Web and voice communications, femtocells can also be used as a solution to enable more value-added services such as remote security, baby monitoring, remote home automation and simpler services like using a phone as a unified home control or multi-line calling.
- Service support – Carriers will need to invest significant sums in the training and equipping of distribution and customer/technical services staff in the support of femtocell solutions. A high return rate will cripple the business model for femtocells and therefore a high priority must be made on quick, easy first-line support.
- Marketing costs – As seen in the Unik and Hot-Spot at Home solutions in France and the United States, there is a need for aggressive and direct marketing solutions in order to ensure

a successful product launch. This is a costly exercise requiring engagement with primary media outlets.

- Infrastructure costs – As well as the cost of the femtocell AP there are also costs associated with controllers, security servers, gateways and other network elements. There may also be some requirement to upgrade some parts of the core network at the same time.
- Connectivity for network elements – One cost aspect of the femtocell solution that is often forgotten is that of connectivity for core elements. These tend to aggregate huge amounts of data traffic and therefore require quite large IP connections that are relatively costly to maintain. This is offset to some extent by the opportunity to offload traffic onto the Internet and home LAN.

3.4.2 Exploring the Economics

Building upon the information in the previous sub-section, it is possible to put together a model of the overall financial impact on a carrier's business. Such a model, developed in (41), is a product of a number of key features such as ARPU, in-home usage, backhaul costs, additional generation of traffic. The model used in this analysis was based upon a Western European WCDMA-based carrier. It includes a number of scenarios based upon subsidisation level, traffic, uptake, cost base, value-added service revenue, backhaul costs, femtocell ASP, data traffic uptake and many other variables. The model takes into account 37 different variables producing 138 different potential outcomes. This section presents the outcomes from the most likely scenario.

Figure 3.4 (42) gives an average picture of the costs involved with a femtocell installation for a typical Iuh-based WCDMA roll out. The model is based on a $128 femtocell wholesale cost with a 100% subsidisation model.

When an average uptake model is based on a moderate uptake in a carrier's base – in the region of 30% – it is possible to create a total cost picture for a number of different scenarios. Figure 3.5(42) provides an insight into the potential total costs for a femtocell solution depending on the cost and level of subsidisation.

On the flip side of these costs there are also significant revenue opportunities for femtocell solutions to spur. Figure 3.6 (42) illustrates the average revenue per annum in US dollars that can be expected for a carrier from a model that is based upon a subscription revenue of $20 a month with a 100% subsidisation level on a femtocell with a wholesale value of $128 per unit.

As well as the revenues accrued from a femtocell solution, a femtocell solution also provides significant saving on OPEX especially in the form of backhaul cost reduction, which is one of the major costs a carrier faces, especially in the face of significant growth in data traffic. Figure 3.7 (42) shows a total picture of average OPEX saving a carrier can expect from reduced backhaul requirements. It must be recognised that this is more applicable to those carriers that rely on leased lines for backhaul transport; those that rely upon microwave transport are highly unlikely to see much benefit in this.

When all of these factors are put together, it is possible to determine the total financial conditions that are relevant to an average femtocell solution. Figure 3.8 (42) shows a number of potential revenue per annum perspectives based on a $128 femtocell average selling price. The worst-case scenario represents a situation where the net effect of the introduction of the femtocell position does not result in any net positive impact on revenues in the form of service revenues and subscription revenues.

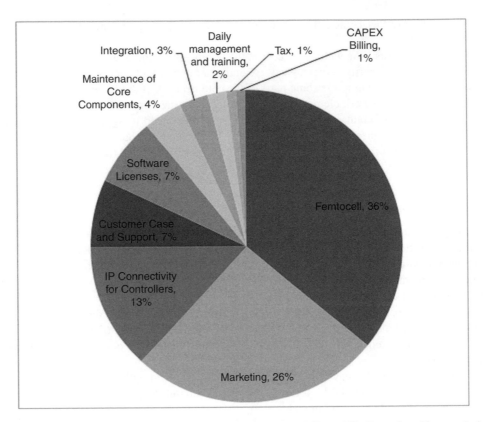

Figure 3.4   Average cost breakdown for a femtocell deployment. from (42). *Reproduced by permission of ABI Research Ltd*

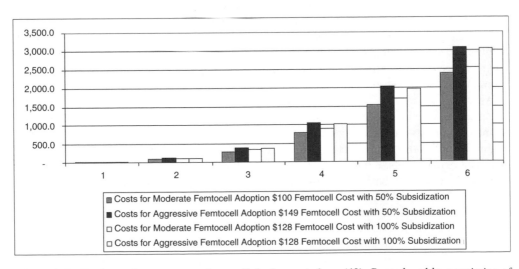

Figure 3.5   Total cost for an average femtocell deployment, from (42). *Reproduced by permission of ABI Research Ltd*

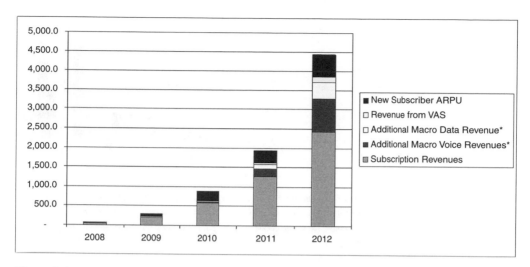

Figure 3.6   Average revenues from a femtocell deployment, from (42). *Reproduced by permission of ABI Research Ltd*

Figure 3.8 clearly shows that, based on the parameters used here, a femtocell solution deployed by a carrier is not a quick-hit solution. It is something that should require a 2–4 year window before the solution breaks even. There is significant investment needed up front in the form of network architecture adaptation, controllers and testing. Real financial positives in terms of large revenue growth will not be seen until five or six years after the initial consideration of the solution. Carriers must view their investments in this light, especially in view of the fact that many other strategic initiatives by carriers take significantly more than 2–4 years before they turn net positive on revenues.

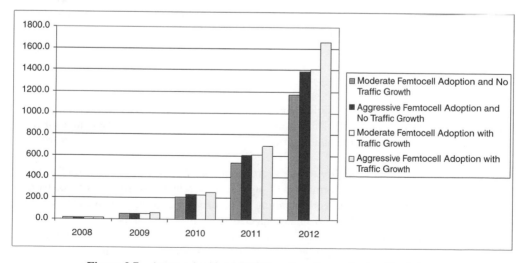

Figure 3.7   Average backhaul OPEX savings from a femtocell solution

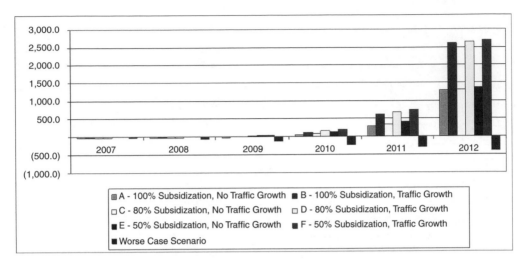

Figure 3.8 An average potential femtocell deployment financial forecast

However, at this point it is important to note that there could be a potential for significantly increasing this projected financial growth if a carrier fully embraces a more progressive service base rather than concentrate on emulating current services.

3.5 Air Interface Choices

It is important to consider the merits of the various air interfaces which may be selected for femtocells and particularly whether 2G-based femtocells (e.g. GSM) or 3G (e.g. WCDMA-based) femtocells will encompass the majority of the market and the opportunity for next-generation femtocells (e.g. LTE or WiMAX). The points below (36) outline how the technologies fit together in terms of competitive offerings.

3.5.1 GSM Advantages

- Handset silicon reuse – It is possible to reuse up to 80% of a handset's chipset design in the production of a GSM femtocell. This allows aggressive price points to be met since up-front development investment is low and femtocells can piggyback price declines, due to volume in the GSM handset arena.
- Frequency planning – The ability to use spare frequency that has been set aside for either guard band or has remained dormant will significantly reduce the impact of macro network interference and make network planning significantly easier than for WCDMA. Some carriers reportedly have 200 kHz bands of spectrum that they are not using.
- Installed subscriber base – The majority of the world's cellular subscribers are currently on GSM networks. This is probably the largest addressable installed base of subscribers available to femtocell vendors and would offer the highest volumes.

3.5.2 GSM Disadvantages

- Pure voice play – It is clear that the GSM offering is a pure voice (and possibly low-rate data) play and will not stimulate growth in data ARPU and will not differentiate from solutions based on VoWi-Fi.
- Spectral efficiency – Carriers are actively looking to move subscribers from GSM networks to WCDMA in many markets in order to benefit from increased spectral efficiency and the associated reduced network investment. It is also important to consider that spectrum is a finite resource that carriers need to nurture. GSM femtocells do not support this goal.
- Future-proofing – When looking at value-added services that may be placed on femtocells, it is important to future-proof from a technological perspective. This cannot be supported by GSM femtocells.

3.5.3 WCDMA Advantages

- Data ARPU uplift – Carriers continually say that they see WCDMA femtocells as a major way of changing user behaviour and enabling subscribers to build affinity with complex data services in the home that will translate into use on the macro network at a higher price.
- Differentiation – The fact that innovative services can be layered onto a WCDMA femtocell, mostly thanks to the higher bandwidth, means that the stickiness of a carrier's bundle can be far higher than just a 'me-too' offering.
- Maximising returns on spectrum investment – Carriers have invested vast sums in 3G licenses and their networks are not returning profits to date. Carriers need to get payback for outlay and WCDMA femtocells can be used as an incentive to migrate.

3.5.4 WCDMA Disadvantages

- Interference – It is far more difficult to accommodate interference in WCDMA networks and carriers are cautious about disrupting a well-oiled macro network operation through the introduction of many smaller femtocells, however, the considerations described in Chapter 4 suggest that this can be overcome with positive net impacts on capacity and performance.
- Cost – WCDMA femtocells are more costly to develop than GSM, due to the reduced reuse of silicon when compared to a handset (only 15% in WCDMA). In addition, the components are more complex and, thus, the raw materials are more expensive. The handsets are more costly and if carriers are going to offer family-type packages they will need to replace some of the handsets since, invariably, some users will still be on GSM.

3.5.5 Conclusions

Although there are benefits from using both 2G and 3G approaches, it appears that carrier attitude is swinging firmly towards more advanced technologies – WCDMA, in particular. It is likely that there will be room for both types of femtocells, thanks to, for example, the low-power GSM licence awards in the UK market (see Chapter 9 for more detail) and the fact that there are carriers with GSM-only networks. It could be argued that, once FMC solutions take hold, it would be sensible for carriers without large investments in WCDMA to enter the market via GSM, due to cost, ease and time-to-market advantages. It is also likely that GSM will continue to play an important role in developing markets if challenges associated with femtocell backhaul can be overcome.

It is also a distinct possibility that the market will begin to segment at an early stage and carriers may want to employ GSM solutions for the most cost-sensitive segments of their subscriber base and use WCDMA for the middle to premium end and price accordingly.

3.5.6 HSDPA, HSUPA and HSPA+

It is interesting to note that carriers are already looking to future-proof devices to make the upgrade path from WCDMA to HSDPA and onwards as smooth as possible. Most vendors will be coming to market from the outset with an HSDPA-enabled femtocell. When considering the upgrade process, carriers are looking for a product that ideally will allow for over-the-air (or IP network) software upgrade.

It is still unclear how HSPA+ upgrades will be facilitated or if there will be the need to plan for some limited equipment swap to encompass MIMO and special diversity requirements that are likely to be an inherent part of the HSPA+ format. It may be that the higher geometry factors for femtocells make this less relevant than for the macrocell environment.

It is important to consider why carriers are so interested in achieving a definable upgrade path. First and foremost, the network in a home or building must reflect the wider network in order to provide a ubiquitous service and also one that is a positive experience in the home (the complete opposite of today's environment). Having HSUPA in the mobile environment and WCDMA only in the home will not work: the disconnect in service delivery will seriously inhibit ARPU growth.

It is also important to consider that greater bandwidth, especially symmetrical bandwidth, will be critical in supporting newer services to handsets. Since users consume at least 70% of their mobile data indoors, it is imperative that the experience be as near perfect as possible.

3.6 Product Feature Sets

As noted above, the feature set requirements are in a state of fluidity (36) with carriers matching feature requirements with service goals and tempering that with cost constraints. It seems certain that some carriers will not be able to come to market with what they consider ideal solutions because of the need to make financial compromises. We are likely to see a more organic growth in product features dictated by affordability and also carrier service mapping.

3.6.1 Stand-alone

The most common approach for early deployments, especially in the WCDMA sphere, is to provide a stand-alone product that simply incorporates the cellular radio aspects of a femtocell. The major reason is the need to meet the challenging price points mooted by carriers. However, this approach does have some important limitations including:

- Potential for service disruption from outside the femtocell – broadband gateway issues.
- Reduction in plug-and-play capability – small numbers of user installations, no matter how few, will dilute the attractiveness of the service solution.

- Support overheads – these products are likely to be low-margin efforts and any significant requirements in terms of product support or box swaps will seriously erode margin. Margin-concerned carriers will look unfavourably at the prospect of support calls caused by third-party equipment.
- Service limitations – this feature set is designed to support a limited fixed mobile convergence play and will not be suitable for additional services such as IPTV and other solutions without a full box swap.
- Home clutter – users do not want extra boxes in their homes. It is likely that they will also need a Wi-Fi access point or ADSL gateway and this is less attractive for subscribers.

Carriers are faced with the decision of balancing the obvious cost benefits of this approach with the potential risks to margins involved with the possibility of vast support requirements. It is likely that this solution will be a good go-to-market approach in a cost-optimised environment, but the carrier needs specific planning and organisational infrastructure in place in order to support the potential impacts of support requirements.

3.6.2 Broadband Gateway

The next step in terms of feature-set enablement is the addition of a broadband gateway so the entire signal chain is routed through equipment that the carriers control, resulting in less chance of support issues caused by equipment from a third party. This type of solution will not significantly increase the number of boxes needed in the home and will facilitate a much smoother plug-and-play offering.

On the downside, there is still the issue of cost. Adding a gateway to the product will add to the cost and, when margins are in single digits, the $15 or so expended on the router and gateway could significantly erode any profit. Moreover, the incorporation of this type of product will probably include another design aspect that will impact on device cost in the development phase.

3.6.3 Wi-Fi Access Point

Adding a Wi-Fi access point is contentious and, on face value, seems a little curious when one of the major motivations of many carriers is to counter the effect and to compete with Wi-Fi-based voice services. However, there are some clear advantages to having Wi-Fi in the package and they are readily identified by carrier RFPs, especially in Europe. First, this reduces the number of boxes needed in the home and, second, the inclusion of Wi-Fi allows additional services to be layered such as the distribution of content to devices outside the handset or general Web surfing on non-handset devices. More important, the inclusion of Wi-Fi is a convenience factor that is somewhat seen as a tick-box feature that will be looked on favourably by consumers.

3.6.4 TV Set-Top Box

The notion of a femtocell that includes a set-top box is more forward looking than other proposals. This final feature provides a device for a full quadruple-play solution and the obvious attractiveness of this product is boundless. However, this may take some time to accomplish and depends significantly on the success of IPTV as a market-proven solution as well as femtocells. Any shortfall for either technology will negate the prospect for this solution and may lead to other products such as satellite or broadcast tuners being incorporated into the offering.

3.6.5 Video Distribution Mechanisms

Increasing the complexity of a solution with an added IPTV capability gets the content from the access point to the point of viewing. A second box will likely be needed at the point of viewing to control the programme menu and interface with the TV. The home network part of the bundle will be important in making this a viable solution. This may open the door to technologies such as 802.11n and UWB as wireless transport mechanisms that can bridge the gap between the femtocell set-top box and the point of viewing. Alternatively it could be that technologies such as LTE and WiMAX could ultimately fill this role.

3.6.6 Segmentation

The discussion of feature sets in this section indicates that there is an enormous chance that the market will segment early in order to tailor those features to meet cost considerations. The real question regarding segmentation is not if but when. The indications are that segmentation will occur rapidly after introduction.

Femtocell vendors will have a significant advantage if they produce mother products where a modular approach could be used to layer in or take out functionality, enabling carriers to segment their product bases.

3.7 Additional Considerations

Beyond the core concept of a residential femtocell, other classes of device, which build on femtocell technology, can be envisaged, namely:

- enterprise femtocells;
- super-femtocell and outdoor femtocell deployments.

3.7.1 Enterprise Femtocells

Although femtocells have been focused initially on the consumer market, it is also becoming increasingly apparent that femtocells also have considerable potential as a solution that is applicable to the enterprise space for improved indoor coverage. When looking at the enterprise in-building space, it is clear that there is a significant solution gap in the sub 5 000 square metre environment and this is the market that is being targeted by carriers.

When looking at the prospects for femtocells in the enterprise setting there are two schools of thought. The first is that consumer-grade products should be used to provide an overlapping coverage map within a building as this solution could benefit from the positive economics seen in the consumer market. The second is to provide a scaled-up option more akin to a picocell that can cover larger areas and accommodate more concurrent users on its footprint.

No matter what option is adopted it is clear that there are some fundamental issues that must be adhered to in order to make an enterprise solution workable. The major factor in these issues is the need for zero-touch installation. If radio planning is required, then the cost of putting in a femtocell-based solution compares unfavourably to distributed antenna systems and picocell solutions. This seems to indicate that a small enterprise environment will be the most likely market for enterprise femtocells.

3.7.2 Super-Femtocells and Outdoor Femtocells

Recent attention has been focused on the introduction of super-femtocells (another name for the class 2 femtocells described in Section 1.4) that incorporate higher capacity for users and greater radio footprints. It now becomes very difficult to differentiate between a picocell and a femtocell. This prospect for super-femtocells seem to be very much a secondary market behind the drive to the introduction of consumer-grade femtocells, but one which has some potential for the future, perhaps even an opportunity for the rebirth of the 3G picocell market.

One of the major applications for these super-femtocells is for outdoor usage in hotspot locations. Although this provides a very interesting concept for carriers where good coverage is difficult to attain, it does have some significant demands in terms of aspects like multi-band operation. It could be argued that there is a role for consumer-grade femtocells in outdoor locations and there are also opportunities for femtocells in enterprises like shops and hospitals where location and presence are important to support applications like advertising.

3.8 Adoption Forecasts and Volumes

This section provides an understanding of the forecasted potential for the femtocell market. At this point, it is important to note that the femtocell market is one that is best characterised as being high risk, high reward. There is no doubt that there is huge potential in the market, but this comes at significant cost to develop products and invest in services.

3.8.1 Methodology

A dual approach was used in the forecasting process to assess the femtocell market. The total available market was produced by taking both a bottom-up and top-down tactic. The new product diffusion model and Bass curve analysis were used to calculate the top-down model, and more details on the process can be seen in Section 3.8.1.1 onwards. The top-down model was then compared to a bottom-up analysis that was compiled by taking each of the different carriers who are looking at the femtocell solution and applying a penetration rate into their subscriber bases.

Further granularity of the forecasts was provided by applying information gathered during the interview phase of this research study in order to model aspects such as feature sets, technology blend and network interface.

3.8.1.1 The New Product Diffusion Model

The growth in penetration of different handset accessories has the characteristics of a 'diffusion model', especially now that the technology has transitioned from something promoted by

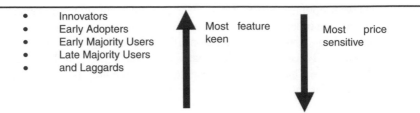

- Innovators
- Early Adopters
- Early Majority Users
- Late Majority Users
- and Laggards

Most feature keen

Most price sensitive

Figure 3.9   Elements of diffusion model of product adoption

manufacturers to products actively demanded by consumers. The model relies on potential users recognising that they have a 'need' or 'desire' that is not being fulfilled; being aware, or being made aware, of a product or service that meets those 'needs' and then being able to afford to buy that product or service.

Research into the diffusion adoption process was pioneered by two key individuals, Everett M. Rogers and Frank Bass. Rogers' research in the 1960s defined the diffusion process into the elements shown in Figure 3.9.

The two major influences of need and awareness result in the net subscriber adoption profile shown in Figure 3.10.

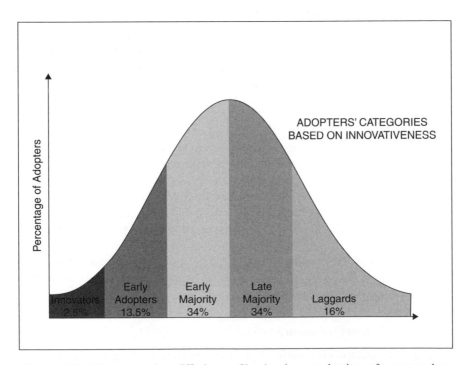

Figure 3.10   The new product diffusion profile, showing net adoptions of a new product

Frank Bass built on this and other research and proposed a model formula that could algebraically describe the new product diffusion profile.

$$N_2 = N_t - 1 + P\left(m - N_t - 1\right) + qN_t - \frac{1}{m\left(m - N_t - 1\right)}$$

N_t indicates the number of people using a given technology or the penetration of a given technology at a given time, t.

Within the formula there are three guiding coefficients:

- m: the ultimate 'market potential' coefficient, that is the maximum attainable ceiling. Usually expressed as a percentage, m generally refers to the addressable market. In the case of cable TV, it could relate to the percentage of homes passed (e.g. 50%) or the attainable cellular adoption market (e.g. 85%).
- P: this can be seen as the 'publicity' coefficient. It is also known as the coefficient of external influence or innovation coefficient. It represents the influence of marketing and publicising of commercial enterprises to drive adoption.
- q: this can be seen as the 'questioner on the grapevine' coefficient. It is also known as the coefficient of internal influence or the coefficient of imitation. This reflects the fact that 'word-of-mouth' recommendations from an existing end-user to a potential end-user can be highly influential.

3.8.1.2 Deriving the Growth Coefficients

The size of m can be derived principally from careful analysis of the market's demographics and levels of disposable income, although other factors such as levels of unemployment, and the percentage level of active subscriptions can also define m.

P and q are derived through a regression analysis of historical data points. The derivation process is complex. Where a particular technology has a track record (i.e. a time series of data points), that track record can be used for the regression analysis. If no track record exists, an analogical approach can be used.

An 'analogical forecast,' or forecast by analogy, involves identifying an existing product with similar features and price characteristics as the new product in question. Historical data is gathered for the analogical example and a regression analysis carried out to acquire the P and q values. These P and q values can then be used in the new product diffusion formula for the product being forecasted.

3.8.1.3 The New Product Conundrum

The derived growth coefficients reflect the adoption profile of an existing product but it is often the case that new products diffuse faster and appeal to a wider audience than an existing product. This can be seen in the adoption profiles of:

- cellular subscriptions compared to fixed-line phones;
- DVD players versus VHS recorders;
- the Internet versus the French Minitel;

- TV versus radio;
- the telephone compared to the telegraph;
- fax machines versus telex, and so on.

While this is true over the complete life cycle of a new product, it does not necessarily take into account lead-in times when a service or product is launched in a limited fashion due to restricted product capacity, insufficient economies of scale, an inappropriately positioned product that does not appeal to the mass market, or substitutions by a competing technology that takes subscriptions away from that emerging technology.

3.8.1.4 Using Analogical Products as a Baseline

Forecasts for the femtocell product use the growth coefficients of an analogical product to create a baseline forecast for the new product. Growth profiles for existing markets were analysed. For each region, five baseline forecast profiles were compiled from regression analyses to represent the highest-growth to the lowest-growth scenarios.

To sum up, growth coefficients of the established end-product markets combined with a judged guidance attained from interviews were used in order to forecast growth in newer product segments where there was no historical data.

Once sufficient reliable and accurate data is received for the new product, it is possible to use that data to generate growth coefficients and revised addressable markets.

3.8.2 Forecasts

Table 3.4 gives an overview of the resulting forecasts for the femtocell market. The table clearly shows that:

- The market will be dominated by consumer products in the short to medium term.
- In terms of enterprise solutions, the small and medium enterprise will take the majority of the enterprise segment due to the need to keep a zero-touch model and the fact that larger enterprise locations would not be able to be supported without radio planning.
- 2009 will see a ramp up in volumes but the real transition to a mass-market product will not occur until 2010.
- There will be a spread of shipments in terms of a product feature set. The market will initially be dominated by stand-alone products due to cost but will soon integrate and be integrated into other solutions.
- The market will be dominated by WCDMA deployments.
- There will be notable profitable niches in CDMA, GSM, LTE and WiMAX markets in particular.
- Multimodality will not make a significant market in the short term due to the drive to reduce cost. It is likely that multimodality will become more apparent as costs reduce and the initial market begins to spread outside its current footprint.
- No particular geographical region will have much ascendancy in terms of volume shipments. The North American market leads initially, but there are a number of prominent carriers in other regions that are on the brink of deployment and this should ensure a fairly even spread between the global regions.

Table 3.4 Femtocell market forecasts

Femtocell Forecasts

World Markets, Forecast: 2006 to 2012

Total Access Point Market	Units	2006	2007	2008	2009	2010	2011	2012	CAGR (09–12)
Access Points Shipped	(Millions)	0.00	0.00	0.12	2.83	9.86	19.37	32.87	127%
Associated Revenue	($ Millions)	–	0.4	16.3	326.7	1,030.8	2,128.1	3,661.6	124%

Air Interface Technology	Units	2006	2007	2008	2009	2010	2011	2012	CAGR (09–12)
GSM/GPRS/ EDGE	(Millions)	0.0	0.0	0.0	0.2	0.8	1.9	3.0	130%
WCDMA/ HSDPA	(Millions)	0.0	0.0	0.1	1.8	6.4	12.4	21.5	128%
Multimode GSM/ WCDMA	(Millions)	0.0	0.0	0.0	0.1	0.5	1.3	3.0	293%
CDMA	(Millions)	0.0	0.0	0.0	0.6	1.8	2.9	4.1	87%
WIMAX	(Millions)	0.0	0.0	0.0	0.1	0.4	0.9	1.4	148%
Multimode CDMA/ WIMAX	(Millions)	0.0	0.0	0.0	0.0	0.0	0.0	0.0	–

Feature Support	Units	2006	2007	2008	2009	2010	2011	2012	CAGR (09–12)
Standalone Femtocell	(Millions)	0.0	0.0	0.1	2.0	6.5	6.6	7.1	52%
Femto/ADSL Gateway	(Millions)	0.0	0.0	0.0	0.7	2.8	6.4	9.5	140%
Femto/ADSL G'way/ Wi-Fi AP	(Millions)	0.0	0.0	0.0	0.1	0.5	6.3	15.6	387%
Femtocell/ ADSL G'way/Wi-Fi AP/IPTV STB	(Millions)	0.0	0.0	0.0	0.0	0.0	0.0	0.6	–

Architecture	Units	2006	2007	2008	2009	2010	2011	2012	CAGR (09–12)
RAN	(Millions)	0.0	0.0	0.0	0.1	0.4	0.6	0.7	67%
Split RNC	(Millions)	0.0	0.0	0.1	2.1	6.9	12.6	19.7	112%
All IP	(Millions)	0.0	0.0	0.0	0.6	2.6	6.2	12.5	172%

(Continued)

Table 3.4   Femtocell market forecasts (*Continued*)

Region	Units	2006	2007	2008	2009	2010	2011	2012	CAGR (09–12)
Asia/PAC	(Millions)	0.0	0.0	0.0	0.7	1.9	3.9	6.4	110%
North America	(Millions)	0.0	0.0	0.0	1.0	3.2	5.8	9.0	106%
Western Europe	(Millions)	0.0	0.0	0.1	1.0	4.6	9.1	15.0	145%
ROW	(Millions)	0.0	0.0	0.0	0.1	0.2	0.6	2.4	–
Customer Type	**Units**	**2006**	**2007**	**2008**	**2009**	**2010**	**2011**	**2012**	**CAGR (09–12)**
Consumer	(Millions)	0.0	0.0	0.1	2.7	9.3	18.0	31.1	126%
SME	(Millions)	0.0	0.0	0.0	0.1	0.5	1.4	1.6	144%
Large Enterprise	(Millions)	0.0	0.0	0.0	0.0	0.0	0.0	0.0	–

Source: ABI Reserach

3.9 Conclusions

In conclusion it is clear that the femtocell market is one that is beset by significant challenges, but also has massive potential opportunities. There is a huge need for significant investment to produce solutions, and there are no guarantees over the success of the solution from a technical and financial standpoint. However, femtocells provide an opportunity to remain competitive in rapidly changing markets. They provide lucrative revenue streams at the same time as offering the promise of helping to support a more economic network operation.

It is very much the case that the considerable promise surrounding this market is providing enough motivation to overrule the potential challenges that are being faced. The carrier community is not attempting to go headlong at this market; it is approaching the challenge cautiously. The femtocell market moved into a period of significant operator activity in 2009 though at low initial volumes, and 2010 will see the market move to a more commercial and open launch of networks, dominated by WCDMA- and CDMA-based stand-alone products.

The market will then evolve quickly into one that is characterised by a number of product tiers and embedded trends, i.e. femto functionality in set-top boxes. The femtocell solution will quickly spread into all major global regions as well as different air-interface technologies such as WiMAX and LTE. The femto offering will transition from one that is dominated by consumer offerings to one that includes a pervasive role in the small and medium enterprise sector.

To sum up: the market should provide a significant opportunity for all, as long as offerings are based on a sound service-driven business model that leverages a well-tested hardware platform.

4

Radio Issues for Femtocells

Simon Saunders

4.1 Introduction

One of the defining characteristics of a femtocell is its use of licensed spectrum. Typically this spectrum is already in use for delivering services using the existing macrocells. Some of the individual frequency channels may be unused by macrocells where operators have sufficient spectrum, but this is not generally the case, so operators need to have confidence that femtocells can operate without creating harmful interference to the existing network, even when deployed entirely by the end-user.

Over the air, femtocells produce identical signals to those which would be produced by a conventional base station. However, there are three respects in which radio issues may differ for femtocells compared with base stations:

- The required coverage area is deliberately limited to the area of a house or small office associated with a given user group.
- Interference between femtocells and macrocells is controlled via entirely automated means rather than via manual planning.
- The cost of femtocells must be minimised, so radio specifications which drive excessive cost without significant performance benefits must be avoided.

This chapter concentrates on the operation of femtocells in wideband CDMA (WCDMA) systems, since these present some of the most challenging cases as well as being candidates for early deployments of femtocells. However, the same general principles and outcomes apply to CDMA systems (e.g. cdma2000) and to OFDMA systems (e.g. LTE and WiMAX).

4.2 Spectrum Scenarios

In WCDMA deployments, each operator typically has only two or three distinct frequencies available and many of these are in bands around 2 GHz, including the 1900 and 2100 MHz

Femtocells: Opportunities and Challenges for Business and Technology   Simon R. Saunders, Stuart Carlaw, Andrea Giustina,
Ravi Raj Bhat, V. Srinivasa Rao and Rasa Siegberg   © 2009 John Wiley & Sons, Ltd

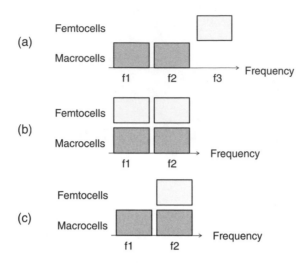

Figure 4.1 Frequency assignment options

bands common in the United States and Europe. Given so few distinct carriers, operators may not have the opportunity to deploy femtocells in dedicated spectrum, so some interference between macrocells and femtocells is possible.

The various cases to be considered are illustrated in Figure 4.1. In case (a), the operator has sufficient carriers to operate femtocells in a distinct carrier to the macrocells, thereby avoiding instances of femto-macro co-channel interference. This may not, however, be the most spectrum-efficient approach, but serves to simplify management and to build confidence in the femtocell deployments. In case (b), the operator has only two carriers available, and decides to use both for a mixture of femtocells and macrocells. This may ultimately provide the highest spectrum efficiency, but requires that the interference mitigation techniques in the femtocells are operating very efficiently and/or that traffic levels are high enough to make this approach essential. This is a relatively unlikely approach in the early stages of femtocell deployment, but may be the end goal when femtocells are extensively deployed. In case (c), the operator chooses to deploy femtocells on only one frequency: thus any macrocell users who experience interference can be handed over to a channel occupied only by macrocells, and thereby extreme cases of interference can be overcome seamlessly. Typically f1 is used as the main coverage frequency for macrocells, with f2 deployed for capacity reasons in particular macrocells, giving rise to more locations where the macrocells and femtocells are well separated. This is a likely approach for early deployments of femtocells and where insufficient spectrum is available to adopt case (a).

Another issue to consider is the use of closed subscriber groups (CSG) versus open access. Open access is the normal manner in which macrocells operate, where all subscribers registered with an operator can access all base stations. Thus all subscribers are subject to power control to ensure they do not transmit at excessive power levels, and all receive service from multiple base stations (soft handover) when they are in areas of overlapping coverage, thereby avoiding interference between macrocells. In femtocells, however, the most common mode of operation will be the CSG approach, where only a limited set of users will be permitted to access the femtocell. Other users will potentially be subject to interference from the femtocells when they are in poor coverage areas from the macrocells but close to the femtocell. They may also cause interference to the femtocell, when they are transmitting at high power to reach a distant macrocell, but are standing close to a femtocell and thereby drowning out a weak femtocell user.

Thus co-channel operation with closed subscriber groups is likely to act as a worst-case scenario when considering interference between femtocells and macrocells and is considered in the calculations in this chapter.

4.3 Propagation in Femtocell Environments

In order to analyse femtocell performance, it is important to determine models of radiowave propagation in typical femtocell environments. Unfortunately, most studies of indoor propagation have focused on larger buildings, such as public buildings and large offices, rather than on the home or small office locations, which will be the main environments for femtocells. There is a large range of variations in the construction styles and arrangements for homes around the world, from densely packed apartment blocks with reinforced concrete and metallised windows, which will give rise to substantial attenuation, to large detached houses of wooden construction, which will exhibit low propagation losses.

However, most studies for the purpose of examining coverage and interference scenarios in general terms have used a model recommended by the ITU-R for the path loss between indoor terminals, known as ITU-R P.1238 (43). The model assumes an aggregate loss through furniture, internal walls and doors represented by a power loss exponent n that depends on the type of building (residential, office, commercial, etc.). Unlike other models, which are site-specific, this method does not require the knowledge of the number of walls between the two terminals. ITU-R P.1238 therefore offers simpler implementation and an ability to generalise across a wide range of buildings. Other models are discussed in detail in Chapter 13 of (14).

The total path loss model (in decibels between isotropic antennas) is:

$$L_{50}(r) = 20 \log f_c + 10n \log r + L_f(n_f) - 28$$

where n is the path loss exponent (Table 4.1) and $L_f(n_f)$ is the floor penetration loss, which varies with the number of penetrated floors n_f (Table 4.2).

Table 4.1 Path loss exponents n for the ITU-R model[a]

Frequency	Environment		
(GHz)	Residential	Office	Commercial
0.9	–	3.3	2.0
1.2–1.3	–	3.2	2.2
1.8–2.0	2.8	3.0	2.2
4.0	–	2.8	2.2
60.0	–	2.2	1.7

[a]The 60 GHz figures apply only within a single room for distances less than around 100 m.

Table 4.2   Floor penetration factors, $L_f(n_f)$[dB] for the ITU-R P.1238 model[a]

Frequency	Environment		
(GHz)	Residential	Office	Commercial
0.9	–	9 (1 floor 19 (2 floors) 24 (3 floors)	–
1.8–2.0	$4\,n_f$	$15 + 4\,(n_f - 1)$	$6 + 3\,(n_f - 1)$

[a]Note that the penetration loss may be overestimated for large numbers of floors, for reasons described in (14).

The path loss $L_{50}(r)$ predicted by this model represents the median path loss at a distance r, i.e. the path loss which is exceeded at 50% of locations at that distance. To convert this into the loss corresponding to a greater proportion of locations, the total loss has to be considered as a sum of this loss and an additional shadow fade margin, L_{FM}. When the model is to be used to assess propagation between an indoor location and a nearby outdoor location, a further additional loss L_W has to be added to represent the outer wall of the building. Thus the total loss is given by:

$$L_T = L_{50}(r) + L_{FM} + L_W = L_{50}(r) + \sigma_L \times F(p) + L_W$$

where σ_L is the location variability characteristic of the indoor environment and $F(p)$ is a coverage confidence factor for the proportion of locations, p, of interest. Note that for points that are distant from the building a different model should be applied to be relevant to an outdoor environment, such as those in Chapters 8 and 12 of (14).

Assuming that the indoor location variability is 6 dB, the fade margin is $L_{FM} = 9.9$ dB for 95% of locations and $L_{FM} = 7.7$ dB for 90% of locations. See Chapter 9 of (14) for further details of these calculations.

The external wall loss is subject to considerable variability according to the building construction, the path geometry and the carrier frequency, but it is common to assume a loss of around $L_W = 10$ dB (44) for residential buildings.

4.4 Coverage

It is desirable first to understand the power levels which are required to deliver adequate femtocell coverage in the absence of co-channel interference, such as in case (a) of Figure 4.1 or when the femtocell is deployed in a remote location where no macrocell coverage is available. The maximum acceptable path loss required to deliver an adequate pilot channel signal quality $\left.\frac{E_c}{N_0}\right|_{femto}$ is then given by:

$$L_{\max} = 10 \cdot \log_{10}\left(\frac{P_{max}}{N_{UE}} \cdot \left(\frac{p_{CPICH}}{\left.\frac{E_c}{N_0}\right|_{femto}} - 1\right)\right) \ (dB)$$

where $P_{\max}$ is the maximum power transmitted by the femtocell (W), N_{UE} is the user equipment receiver noise power and p_{CPICH} is the proportion of the femtocell power allocated to the pilot channel. Figure 4.2 illustrates the calculation of the coverage versus the femtocell transmit

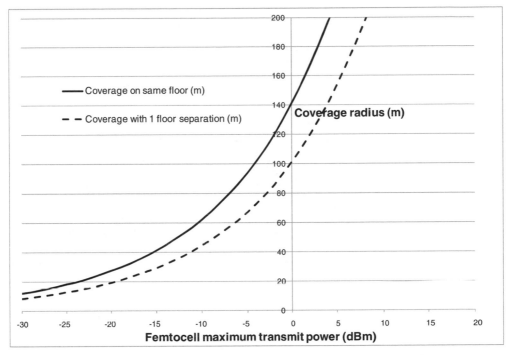

Figure 4.2 Femtocell coverage radius for noise-limited environments, assuming $\left.\frac{E_c}{N_0}\right|_{femto} = -16$ dB, $p_{CPICH} = 0.1$ and that the user equipment noise figure is 7 dB and the fade margin $L_{FM} = 7.7$ dB, giving around 90% availability at the cell edge

power, including both coverage on the same floor and with one floor separation between the femtocell and the user. Coverage levels well in excess of that required for a typical house are achieved even at transmit powers below –10 dBm (0.1 mW). For comparison, a Wi-Fi access point typically transmits at +20 dBm (100 mW), and phones typically transmit at up to +24 dBm (250 mW).

Having established that femtocells are capable of providing very good coverage and high data rates to even large homes in isolated locations (i.e. those where there is negligible signal from co-channel macrocells), we next need to consider the case where the femtocell is operated in the presence of surrounding macrocells.

4.5 Downlink Interference

Why should a femtocell work in the presence of macrocells in the first place? After all, it is supposed to provide extra coverage and capacity using spectrum which is surely already well used by the macrocell network, which has to be built to provide indoor coverage over a wide area. However, as Figure 4.3 illustrates, the femtocell and its users benefit from the insulating effect of losses through the exterior house walls, producing benefit from the same

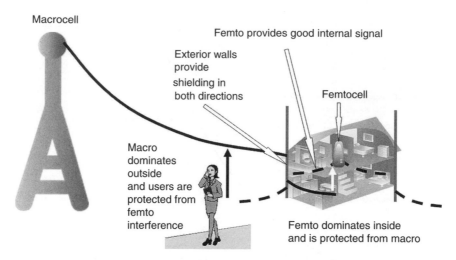

Figure 4.3   Femtocells operate in the presence of the macro network, insulated by the exterior walls of the residence

losses which make it challenging for the macrocells to provide good coverage in the first place.

The femtocell is within the house, where losses due to walls and floors are fairly small, so it can provide coverage throughout a wide area with low power. The femtocell users are protected from interference from the macrocell network by the exterior walls of the building, which cause the femtocell power to diminish rapidly outside.

The same external wall losses also protect nearby macrocell users from femtocell coverage. The femtocell power is only high enough to just serve the internal users, while the macrocell user sees only an attenuated version of the femtocell signal and experiences little degradation, having a strong macrocell signal to access in most cases. The femtocell signals also diminish rapidly with distance outdoors, so any interference which does occur is only felt when in close proximity. Even if some increase in interference is experienced by the macrocell user, the macrocell will simply allocate a little more power to affected users, thereby maintaining their signal quality with little impact on other users.

However, this reference case represents a situation where the scenario is working in the favour of the femtocell. More generally, we need to consider more challenging cases where the macrocell user may be operating at the limit of macrocell coverage, having available only a weak signal and a low tolerance to additional interference and also where the femtocell is deployed in a challenging situation, e.g. near the edge of the house and close to the window, thereby delivering its highest levels of interference to the user. In this situation and without special mitigation techniques being applied, the femtocell can create a 'dead zone' around it, where the service available to the macrocell user is degraded (e.g. poorer speech quality or reduced data rates) or even unavailable.

In order to assess the impact of such dead zones, the path loss between a macrocell user and the femtocell for a given level of service degradation can be expressed as follows (45):

$$L_h = 10 \cdot \log_{10} \frac{a \cdot P_{max}}{ACIR \cdot \left(\dfrac{RSCP_{best\_macro\_CPICH}}{\left. \frac{E_c}{N_0} \right|_{macro}} - RSSI_{macro} \right)}$$

where L_h is the path loss between the macrocell user and the femtocell (in decibels, 0 dBi gain antennas assumed), a is the ratio between the mean and maximum transmit power of the femtocell, P_{max} is the maximum power of the femtocell (watts), $ACIR$ is the adjacent channel interference ratio ($= 1$ for co-channel operation), $RSCP_{best\_macro\_CPICH}$ is the pilot channel code power received by the macro user from the strongest macrocell (watts), $\left. \frac{E_c}{N_0} \right|_{macro}$ is the required signal to interference ratio for the macrocell to achieve adequate quality and $RSSI_{macro}$ is the overall received signal strength from the macrocell (watts). The pilot power is related to the macro signal strength via $\left. \frac{E_c}{N_0} \right|_{no\_femto}$, the signal to interference ratio in the absence of the femtocell, as follows:

$$RSCP_{best\_macro\_CPICH} = RSSI_{macro} + \left. \frac{E_c}{N_0} \right|_{no\_femto}$$

Predictions for the dead-zone size are shown in Figure 4.4 for various femtocell transmit powers. The size of the dead zone varies across the macrocell coverage area, shrinking as the femtocell comes closer to the macrocell as the macrocell user has a dominant signal from the macrocell. For strong macrocell signal strength, the dead zone is only a few metres in radius,

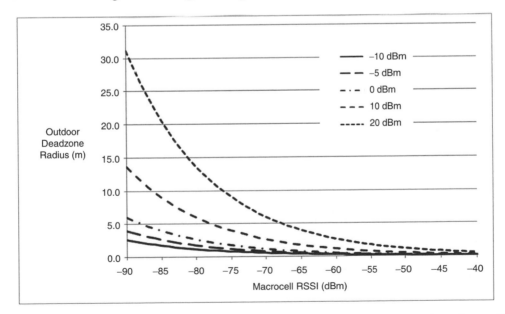

Figure 4.4 Co-channel deadzone radius versus macrocell signal strength for various femtocell maximum transmit powers. Here $L_W = 10$ dB, $L_{FM} = 7.7$ dB, $a = 0.3$, $\left. \frac{E_c}{N_0} \right|_{macro} = -16$ dB and $\left. \frac{E_c}{N_0} \right|_{no\_femto} = -8$ dB

which would anyway only affect users inside the premises. For weaker macrocell signals, the dead-zone radius rises and for higher powers would be excessively large.

In order to avoid dead zones of excessive size, the natural remedy would be for the femtocell to vary its transmit power according to the macrocell signal strength in the local area. But the question then arises as to whether the femtocell signal will then dominate over the macrocell inside the premises sufficiently to deliver a good service. The loss associated with the femtocell coverage in the presence of macrocells can be expressed as follows:

$$L_q = 10 \cdot \log_{10} \left(\frac{P_{max}}{\frac{RSSI_{macro-N_{UE}}}{ACIR} + N_{UE}} \cdot \left(\frac{P_{CPICH}}{\frac{E_c}{N_0}\big|_{HNB}} - 1 \right) \right)$$

If the femtocell maximum transmit power is adjusted to deliver a fixed service area, depending on the level of macrocell interference observed, then the dead-zone size can be minimised. Indeed in this case, since both the femtocell power and the macrocell interference vary in the same proportions, the dead-zone radius is actually independent of the level of macrocell interference, as illustrated in Figure 4.5.

For strong macrocell signal strengths, the required femtocell power may exceed its maximum capability (typically 10–20 dBm). In these situations, some reduction in femtocell radius may be acceptable, or any affected macrocell users may be handed over to a channel clear of femtocells, avoiding the dead-zone interference.

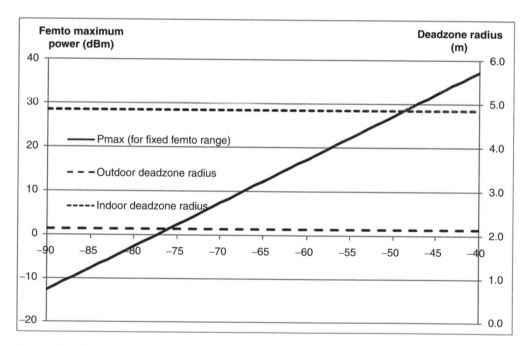

Figure 4.5 Femtocell transmit power required to deliver a coverage radius of 20 m as a function of the macrocell signal strength, together with the associated deadzone radius for an indoor or outdoor macrocell user

4.6 Interference Challenges and Mitigations

As well as the general downlink dead-zone issue described in the previous section, a number of other interference situations which occur in extreme situations can also be identified. These are explained in Table 4.3, together with their potential impact.

All of these situations have been examined in great detail by a variety of organisations. In particular, see the collection of studies created within 3GPP as part of its work assessing the feasibility of WCDMA femtocells (46) and the white paper published by the Femto Forum (47), which examined particular 'corner cases' in a variety of ways and established the performance in these cases.

Overall these studies showed that there is a clear need for femtocells to implement interference mitigation techniques in order to avoid the occasional extreme cases where interference can occur. The specifics of these are left to individual vendors, but Table 4.4 lists some of the key techniques which have been examined.

Using a combination of these techniques, the studies in (46) and (47) have shown that, when comparing networks having only macrocells with those having a mixture of femtocells and macrocells, the net effect is to improve overall performance for both femtocell and macrocell users, while significantly increasing overall network capacity. In the small number of extreme situations where isolated interference instances can exist, the network can detect these occurrences and manage calls appropriately to avoid deterioration of the experience of users as a whole across the network.

Guidance on the relevant issues and mitigation techniques is also now provided in the 3GPP standard in a technical report (48). Specific algorithms for interference mitigation are not included, allowing vendors to continue to innovate in ensuring high performance in this area.

Table 4.3 Potential interference scenarios for femtocells

Potential issue	Potential impact
Downlink power from femtocells with closed subscriber group causes interference to macrocell user	Macrocell user experiences degraded service and potential loss of service
Femtocell user at edge of femto coverage transmits at high power, causing noise rise to nearby macrocells	Macrocell users at edge of coverage experience degraded service
Macrocell user close to femtocell but far from macrocell operates at high power, causing interference and potentially receiver blocking to femtocell	Femtocell users experience degraded coverage and service
Femtocell user at edge of coverage of femtocell 1 but close to femtocell 2	User experiences degraded downlink service due to interference from femtocell 1 and transmits at high power, degrading uplink service for users of femtocell 2

Table 4.4 Interference mitigation techniques for femtocells

Mitigation technique	Explanation and usage
Channel assignment	The network assigns users who are not part of the femtocell subscriber group to the most appropriate channel. Users who are on the macrocell can avoid femtocell dead zones. Femtocell users can be assigned different channels to avoid interference in overlapping coverage areas
Downlink power management (i.e. automatic gain control)	The femtocell transmit power is adjusted to give an appropriate trade-off between coverage and interference at a given location. This may be done using direct measurements of both the uplink and downlink channels and using measurements taken by both femtocells and user equipments to provide enhanced accuracy. The levels may also be varied over time in response to changing conditions
Power capping of user maximum transmit power	The femtocell sends a broadcast message to mobiles in its coverage to ensure a given maximum transmit power is never exceeded. As users leave the femtocell coverage, they are thus prevented from causing excessive uplink interference to the macrocells and the coverage area
Dynamic receiver gain management	An adaptive attenuation level is included in the femtocell receiver to reduce its gain when a strong co-channel (or adjacent channel) mobile is nearby, keeping the receiver operating within its linear dynamic range and avoiding blocking while still providing sufficient sensitivity to detect mobile at the edge of femtocell coverage
Broad dynamic range specification and testing	Since mobiles can approach very close to femtocells, the strong signal-handling ability of the femtocell should be verified in conformance testing

4.7 Femtocell-to-Femtocell Interference

We have established that femtocells with the right interference mitigation techniques are capable of operating co-channel with macrocells to deliver enhanced performance and capacity, with means of avoiding residual instances of interference in the extreme and unusual cases where it still occurs.

However, when femtocell deployments become very dense, e.g. in adjoining apartments in an apartment block or in a dense office building, it becomes relevant to consider whether the same techniques are also capable of dealing with potential co-channel interference between individual femtocells.

By way of example, Figure 4.6 shows the layout of a real-world trial on one floor of an office building. Twenty femtocells, all on the same channel, are deployed to cover individual offices with areas between 12.5 m² and 50 m². The floor has a total area of 700 m² and the internal walls dividing the offices are thin and of light construction. As well as the femtocells,

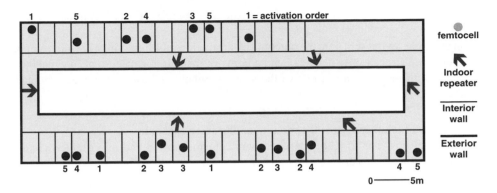

Figure 4.6 Layout for trial of dense femtocell deployment trial. The black circles are the femtocells and the white triangles are indoor repeaters. *Reproduced by permission of Ubiquisys Ltd*

there is co-channel macrocell interference and a series of indoor repeaters, contributing to an interference level in excess of –65 dBm. Extensive user trials and measurements of the system were conducted with assessments made against a number of performance goals. The mobile operator feedback is assessed against each of these goals in Table 4.5. The femtocells are found to adapt well to deliver coverage over the desired areas with good performance despite the presence of very strong co-channel interference, with each delivering a full carrier of HSDPA capacity, resulting in very high traffic density and spectrum efficiency.

4.8 System-Level Performance

Our analysis so far has addressed mainly femtocell performance when considering the environment immediately around a given femtocell. In order to assess the overall performance of a network in which femtocells are deployed, it is important to conduct complete

Table 4.5 Operator feedback from dense femtocell trial

Performance goal	Operator feedback
Can femtocell autoconfiguration operate successfully in the presence of strong macrocell interference?	'Femtocell is auto-adapting well to macro environment (strong to very strong macro signals)'
Can femtocells provide adequate coverage in the service areas assigned to them in the presence of strong macrocell interference?	'The femtocells were able to cover the area within the rooms they are placed in with interference levels exceeding –65 dBm'
Does HSDPA perform well?	'HSDPA within the femtocell coverage extremely good'
How much capacity is delivered?	'Even in dense environment and maximum load on neighbouring femtocell, each femtocell is able to support the maximum specified capacity'

system-level investigations, which include realistic distributions of user locations and propagation conditions.

As an example, we present here the outcome of simulations published in (47) which compare a dense urban macrocell-only network with a network including a mixture of macrocells and femtocells. Both networks aim to serve the same population of users. The main simulation parameters are:

- macrocells and femtocells are all operating on the same channel;
- pilot channel power is 10% of the maximum transmit power and all common channels together constitute an overhead of 25% of the maximum;
- one femto user per femtocell;
- femtocell transmit power is adjusted according to the macrocell power measured in the surrounding area.

The results show that, when femtocells are deployed, there are some instances of macrocell users suffering interference from femtocells, but these are balanced by indoor macrocell users who gain service from femtocells, so that the femtocells do not introduce outage overall. In practice, any macrocell users who encounter interference would be switched to a different channel and would avoid outage from the femtocells.

Figure 4.7 shows the system capacity benefit from deploying femtocells. Without femtocells, almost all users get a throughput less than 400 kbps. When femtocells are deployed, more than 70% of users get a throughput greater than 400 kbps. Table 4.6 compares the overall

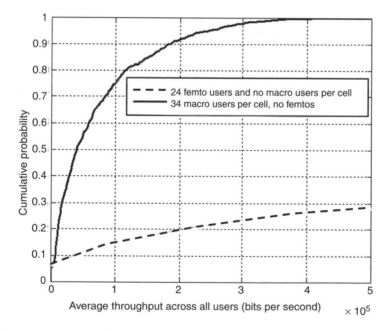

Figure 4.7   Average throughput of users from simulations with and without femtocells. *Reproduced by permission of Femto Forum Ltd*

Table 4.6 Comparison of overall network capacity with and without femtocells

| | Capacity/Mbps | |
	Femto + macro	Macro only
Expected available throughput per user	7.87	0.08
Number of users per cell	34	34
Expected available throughput per macrocell (= network capacity)	267.7	2.6

network capacity, where throughput gains of around 100 times are available to users accessing the femtocells on average and around 10 times in total including the macrocell users. Note that the macrocell users also benefit due to the reduced loading on the macrocells.

In summary, the simulations show that the deployment of femtocells with appropriate interference mitigation techniques not only improves the throughput distribution but also the coverage, assuming multiple carriers are available.

In a second simulation example (49), a system consisting of a realistic mixed macrocell and microcell network as deployed in a dense area of around 2.5 km² in London was simulated. Femtocells were randomly distributed into apartment blocks with multiple femtocells each serving up to three users. The simulation area is illustrated in Figure 4.8. Both uplink and downlink performance was simulated and the femtocells implemented relevant interference mitigation techniques. In total the simulation included 500 indoor users and 600 outdoor users, each consuming a mixture of speech and data traffic at a mixture of speeds at 3 km/h (indoor and outdoor users) and 50 km/h (outdoor users) with 250 femtocells. A closed access,

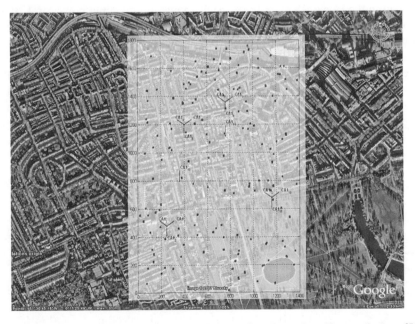

Figure 4.8 Simulation area for downlink and uplink simulation. *Reproduced by permission of Ubiquisys Ltd*

co-channel deployment was assumed. Only users in the same apartment were able to access the femtocell, while all others represented interference.

The simulation increased the volume of traffic with only the macrocells active, until a significant quantity of dropped calls was observed. At this stage 60% of macrocell users had services while only 30% of indoor users did. With the femtocells switched on, 99.9% of the indoor users and 80% of the outdoor users were successfully served. Translating the results into spectrum efficiency, the macro + micro network delivered 1.1 bits per second per hertz per square kilometre, while the combined macro + micro + femto network increased the spectrum efficiency to over 2.5 bits per second per hertz per square kilometre.

4.9 RF Specifications in WCDMA

The 3GPP WCDMA standards is used here as an example of the distinctive RF performance specifications for femtocells, but other standards will incorporate comparable performance.

Over the air, femtocells are essentially identical to existing base stations, allowing existing mobiles to operate without any changes. In the main, the RF performance of femtocells is specified for WCDMA to be the same as the existing 'Local Area BS' class, which is intended for picocells. However, there are areas where some changes to the RF specifications enable reduced cost of implementation where the picocell performance is excessive for femtocell applications or where there are specific interference conditions to be protected against. These differences are specified via a new WCDMA base station class, the 'Home BS' to apply to femtocell scenarios. This was done in document TS 25.104 of Release 8 of the 3GPP standard (50), which was functionally frozen in December 2008. The differences between the Local Area BS and the Home BS are specified below.

Equipment in the Home BS class has a specified maximum output power of +20 dBm (i.e. 100 mW). This is the mean power level for a single carrier at the antenna connector. Note that antenna gain could increase the effective radiated power beyond this value. If transmit diversity or MIMO is applied then the output power is 3 dB less (+17 dBm or 50 mW).

The frequency error is specified as ±0.25 parts per million (ppm), compared with ±0.05 ppm for the Wide Area BS class and ±0.1 ppm for the Medium and Local Area BS classes. The reduced tolerance is allowable because the lower indoor user speeds mean that less allowance needs to be made for doppler shifts, which in turn means that a greater error can be tolerated in the base station without increasing the overall tolerable frequency offset seen by the user equipments.

An additional spectrum emission limit is specified as shown in Tables 4.7 and 4.8. A spurious emission limit of −82 dBm in a measurement bandwidth of 100 kHz is specified.

Table 4.7 Additional spectrum emission limit for Home BS, BS maximum output power $6 \leq P \leq 20$ dBm

Frequency offset of measurement filter −3 dB point, Δf	Frequency offset of measurement filter centre frequency, f_offset	Additional requirement	Measurement bandwidth (Note 2)
12.5 MHz $\leq \Delta f \leq$ Δf_{max}	13 MHz $\leq$ f_offset $<$ f_offset$_{max}$	P - 56 dBm	1 MHz

Table 4.8   Additional spectrum emission limit for Home BS, BS maximum output power $P < 6$ dBm

Frequency offset of measurement filter –3 dB point, Δf	Frequency offset of measurement filter centre frequency, f_offset	Additional requirement	Measurement bandwidth
12.5 MHz $\leq \Delta f \leq$ Δf_{max}	13 MHz $\leq$ f_offset $<$ f_offset$_{max}$	–50 dBm	1 MHz

The receiver reference sensitivity for the Home BS is –107 dBm, the same as for the Local Area BS class, but higher than the Medium and Wide Area BS classes. This maintains a good opportunity to reduce the transmit power needed from mobiles attached to femtocells.

The receiver blocking characteristics are specified as the same as the Local Area BS, but the requirements for the Home BS when co-located with DECT and Wi-Fi are flagged for further study and potential extension of the specification in future releases.

There is a new requirement, applicable only to the Home BS, to adjust the transmit power of the femtocell to provide protection to adjacent channel mobiles which are close to the femtocell but which belong to other operators. The power is set as a balance between the interference level observed on the femtocell channel (excluding the femtocell itself), Io, and the adjacent channel primary channel code power, CPICH Êc, according to Table 4.9.

On the other hand, the Home BS adjacent channel leakage power is permitted to be the higher of the normal adjacent channel leakage ratio (45 dB for first adjacent channel) or –50 dBm/MHz, whichever is higher, recognising that at the lower output powers for femtocells the absolute adjacent channel leakage power is already small, thereby assisting in implementation of the transmitter without excessively degrading adjacent channel performance.

The dynamic range of a Home BS is tested to a higher level than for the Local Area BS, at –39 dBm rather than –59 dBm, ensuring that it can perform well even when mobiles are very close to the femtocell.

The 3GPP performance requirements specify several propagation conditions relating to multipath delay spread due to varying path lengths for reflected and scattered paths and also relating to channel fading rates arising from doppler spread (see (14) for details). The Home BS only has to meet specified performance targets for multipath cases which are 'static', 'case 1' or 'Pedestrian A', relating to a maximum multipath delay of 976 nanoseconds and a maximum fading speed corresponding to 3 km/h in the 2.1 GHz band. This compares with delays of up to 20 000 ns and speeds up to 120 km/h in other channels. This allows the signal

Table 4.9   Home BS output power for adjacent channel operator protection

Input conditions	Output power, Pout (without transmit diversity or MIMO)	Output power, Pout (with transmit diversity or MIMO)
Io > CPICH Êc + 43 dB and CPICH Êc $\geq$ –105dBm	$\leq$ 10 dBm	$\leq$ 7 dBm
Io $\leq$ CPICH Êc + 43 dB and CPICH Êc $\geq$ –105 dBm	$\leq$ max(8 dBm, min(20 dBm, CPICH Êc + 100 dB))	$\leq$ max(5 dBm, min(17 dBm, CPICH Êc + 97 dB))
CPICH Êc < –105 dBm	$\leq$ 20 dBm	$\leq$ 17 dBm

processing complexity of the femtocell receiver to be reduced by avoiding a Rake receiver with a large number of fingers and a large correlator search window.

These RF performance requirements are also reflected in the base station conformance test specifications, which incorporate tests specific to the Home BS class (51).

4.10 Health and Safety Concerns[1]

Femtocells, along with other devices used in the home, are transmitting devices. Although, as we have seen earlier in this chapter, femtocells operate at power levels which are at or below the levels of existing devices such as Wi-Fi access points or DECT-based cordless phones, there can be understandable concern at the operation of these devices in the home by those unfamiliar with the scientific background and the precautionary steps taken by the industry. This section summarises the scientific background and the status of relevant regulatory limits, together with methods of ensuring compliance with these limits.

Femtocells emit very low levels of radiowaves, also known as radio frequency (RF) electromagnetic (EM) fields. In the frequency range of interest for femtocells, electromagnetic radiation is referred to as 'non-ionising radiation' as distinct from the ionising radiation produced by radioactive sources. The energy associated with the quantum packets or *photons* at these frequencies is insufficient to dissociate electrons from atoms, whatever the power density, so the main source of interactions between non-ionising radiation and surrounding human tissue is simple heating. At the levels delivered by femtocells, the heating impact is exceedingly small.

The potential health impact of EM fields has been studied for many years by both civil and military organisations. Numerous independent bodies have commissioned research into such effects and the World Health Organisation has produced guidelines to ensure that this research is conducted according to appropriate standards (52). Scientific expert panels, health agencies and standard-setting organisations around the world regularly review this large and growing body of research. The conclusions from these investigations have been used to set regulatory limits on exposure, which reflect a precautionary principle based on the current state of knowledge. These organisations have all reached the same general scientific conclusion: there are no established health effects from exposure to radiowaves below the applicable limits. Many administrations require equipment manufacturers to ensure that the fields absorbed are below these limits and to quote the values produced by individual equipments under suitable reference conditions.

The limits in question are based on the levels of the EM fields incident on the human body and the associated energy absorption inside the human body (53). The absorption is measured in terms of the *specific absorption rate* (SAR), or the rate of change of incremental energy absorbed by an incremental mass contained in a volume of given density, with units of watts per kilogram, or equivalently milliwatts per gram. The SAR (divided by the specific heat capacity of the material) indicates the instantaneous rate of temperature increase possible in a given region of tissue, although the actual temperature rise depends on the rate at which the heat is conducted away from the region, both directly and via the flow of fluids such as blood.

---

[1] This section is adapted by permission from (14), which provides further technical details.

At the early design stages of a transmitting system, it is possible to make an assessment of the likely SAR via analytical or numerical calculations, although such calculations are complex and not sufficient for final assessment. More usefully, SAR is determined via measurements. A popular method to perform SAR measurements is by logging electric field data in an artificial shape acting as a representation of the human body in normal use, from which SAR distribution and peak averaged SAR can be computed. To do so repeatably, it is important to use an appropriate body shape and dielectric material. Appropriate representation for the human head and other body parts are standardised by relevant committees ((54), (55)). In 2001, the European Committee for Electrotechnical Standardisation (CENELEC) published a series of technical standards detailing how to make SAR measurements and the industry is now publishing SAR values for phones (56), (57), (58), (59). This standard has been harmonised across the EU member states, but compliance remains voluntary, and manufacturers are free to choose any other technical solution that provides compliance with the essential requirements.

After characterising the RF exposure produced through SAR measurements, it is necessary to assess whether this exposure falls beyond acceptable limits. ICNIRP (International Committee on Non-ionising Radiation Protection) is an independent non-governmental scientific organisation, set up by the World Health Organisation and the International Labour Office, responsible for providing guidance and advice on the health hazards of non-ionising radiation exposure (60). In 1999, the European Committee for Electrotechnical Standardisation (CENELEC) endorsed the guidelines set by ICNIRP on exposure reference levels, and recommended that these should form the basis of the European standard (61). Levels based on the ICNIRP recommendations are also adopted in various other regions of the world. In the Americas, the IEEE C95.1-1999 Standard for Safety Levels with Respect to Human Exposure to Radio Frequency Electromagnetic Fields has been adopted as a reference (62). All of these limits are based on a precautionary approach, to be substantially below the levels at which adverse effects due to heating occur.

Table 4.10 shows the basic SAR limits for both the ICNIRP and IEEE standards. Both standards make a clear distinction between general public (uncontrolled environment) and occupational (controlled environment). For the former, people with no knowledge of or no control over their exposure are included, and hence the exposure limits need to be tighter. However, the general public values are often regarded as representing best practice, whoever the affected parties.

The ICNIRP and IEEE standards also establish field strength and power density limits for far-field exposure, as shown in Tables 4.11 and 4.12. Notice the variations in maximum electric field exposure with frequency.

Table 4.10   SAR exposure limits (W/kg) (53)

Standard	Condition	Frequency	Whole body	Local SAR (head and trunk)	Local SAR (limbs)
ICNIRP	Occupational	100 kHz–10 GHz	0.4	10	20
	General public	100 kHz–10 GHz	0.08	2	4
IEEE	Controlled	100 kHz–6 GHz	0.4	8	20
	Uncontrolled	100 kHz–6 GHz	0.08	1.6	4

Table 4.11 ICNIRP reference field strength levels (Vm^{-1})

Standard	Condition	> 10 MHz < 400 MHz	900 MHz	1.8 GHz	> 2 GHz < 300 GHz
ICNIRP	General public	28	41.25	58.3	61
	Occupational	61	90	127.3	137

For low power emissions it is possible that the ICNIRP public exclusion zone is contained entirely within the casing of the femtocell, so that the device is ICNIRP compliant in any possible usage condition, including when the user touches the case of the femtocell. The European standard EN50385 states the following concerning compliance (57):

> If the average power emitted by the base station is less than or equal to 20 mW then the base station is deemed to comply without testing.
>
> If the average power emitted by the base station is more than 20 mW, then E, H or SAR calculations and/or measurements shall be performed according to clause 4. The results of calculations and/or measurements shall be compared directly to the limits.

Hence, provided that the power into the antenna is less than 20 mW in total (i.e. including all transmissions across all bands), compliance with ICNIRP may be assumed. This does not indicate that a femtocell radiating more than 20 mW is not compliant, it simply means that a further assessment of compliance should be performed via SAR testing. Such testing is usually performed by independent laboratories on a representative sample of a device of a given design. There is no need to test every device once approval has been given, as long as there are no relevant changes to the design.

In summary, femtocell vendors are designing their products to fully comply with the guidelines for human exposure to electromagnetic emissions issued by the International Commission on Non-Ionising Radiation Protection (ICNIRP) and other relevant regulatory authorities. The low-power nature of femtocells means that such compliance is well established, so femtocells should create no further concern given the widespread acceptance in the home and office of devices such as cordless phones and Wi-Fi access points.

In order to explain the facts regarding femtocells and health in a form suitable for users of femtocells, a short brochure has been produced by the Femto Forum, the Mobile Manufacturers' Forum and the GSM Association (63).

Table 4.12 IEEE reference electric field and power density levels

Standard	Condition	Electric field strength (Vm^{-1}) > 30M Hz < 300 MHz	Power density ($mWcm^{-2}$) 900 MHz	1.8 GHz	> 15 GHz < 300 GHz
IEEE	Uncontrolled	27.5	0.6	1.2	10
	Controlled	61.4	3	6	10

4.11 Conclusions

In order to deliver the benefits of coverage, capacity and spectrum efficiency in the full range of operating conditions, femtocells must operate reliably in the presence of existing macrocell and microcell networks, which will typically operate within the same spectrum. Several potential interference mechanisms have been identified, which need to be managed to avoid performance degradation in occasional extreme situations. There also exists a variety of automated interference mitigation techniques capable of improving performance in these situations. Numerous studies and increasing numbers of practical measurements have demonstrated that, if the interference mitigation techniques are appropriately implemented, they deliver performance within the network which improves overall performance for all users, including those on both the macrocell and femtocell networks. The result is a very substantial increase in capacity and spectrum efficiency for the mobile network as a whole.

To realise these interference mitigation techniques and to allow femtocells to be produced cost effectively, standards bodies are now incorporating specific RF characteristics, which are appropriate to femtocells, including reduced transmit powers and performance characteristics applicable to the intended domestic and office environments of femtocells.

These low power levels also ensure that femtocells can be produced fully compliant with relevant limits on electromagnetic emissions, similar to existing, well-accepted home and office devices, helping to ensure that the devices are acceptable to users when widely deployed and delivering other benefits such as reduced interference and increased mobile transmit power and battery life.

5

Femtocell Networks and Architectures

Andrea Giustina

5.1 Introduction

In the last two years, technology has matured enough to make it possible to support a high-performance femtocell on a consumer-priced platform (see Chapter 10). Many leading operators evaluating the coverage, capacity and new services opportunities provided by femtocells at a very effective cost (see Section 1.9) have expressed a strong desire to deploy femtocells rapidly. This has driven a number of telecommunication infrastructure vendors and new players to develop femtocell products.

Femtocells are an extension of the operator network into end-user premises and they present particular architectural challenges that must be overcome in order to enable commercial deployments. Furthermore, the speed of innovation has meant that virtually every femtocell vendor initially developed its own architecture, with many of them utilised in trials and early commercial deployments, posing difficult problems for converging these disparate solutions to industry-level standards.

Luckily there is a lot of common ground between the different architectures, providing a good starting point for such standardisation work, and the Femto Forum has played a pivotal role in bringing the industry players together to define how to best approach standardisation across the main radio network technologies (see Chapter 8).

This chapter first addresses the challenges and requirements of the femto architecture, then defines the common femto design principles found across different technologies, with the functional split between the new network elements and the interfacing with the rest of the core network elements. It then reviews the architectures and, where available, emerging standards specific to UMTS, CDMA, WiMAX, GSM and LTE.

Femtocells: Opportunities and Challenges for Business and Technology   Simon R. Saunders, Stuart Carlaw, Andrea Giustina,
Ravi Raj Bhat, V. Srinivasa Rao and Rasa Siegberg   © 2009 John Wiley & Sons, Ltd

5.2 Challenges

The existing mobile network architectures, especially for GSM, UMTS and CDMA technologies, but also for LTE and to some extent WiMAX, were defined for large-scale, hierarchical network deployments with network elements that, apart from the end-user terminal access, are typically installed in secure premises, use dedicated high-performance links to interconnect and have proprietary element management systems.

Femtocells pose a quite different set of challenges to these architectures, the main ones being:

- Femtocells can be deployed in potentially millions of units in a single network, hence requiring a quite different scalability: up to 1000 times more femtocells than macrocells can be deployed in the same network. Typically mobile networks allow for up to a few hundred macrocells to connect to the next level up of the hierarchy, so this kind of scalability is not sufficient for femtocell deployments.
- Femtocells are mostly deployed in end-user premises, hence in inherently insecure environments requiring hardened units but also an architecture that enables the sufficient level of security, protecting both the end-user and the mobile networks from security risks.
- Femtocells connect over shared broadband IP links with variable performance, so must adopt techniques that are more typical of Internet and VoIP applications compared to those typical in mobile networks.
- Femtocells must integrate as transparently as possible into the existing operational processes of mobile network operators.

Femtocells must be remotely managed in an efficient way for the typical scale of deployments and for multi-vendor deployments. Typical mobile network management is intensive, requesting a tight coupling between the network elements and their management system, works over dedicated secure links and is vendor specific (aka 'proprietary'). Femtocells require an open-standard management system that works in non-real-time and with low-intensity interactions with the management system, i.e. femtocells must self-manage where possible, although always within parameter ranges and policies which are set by the operator.

The existing network architectures are thus unable to cope with a commercial femto deployment, even though in a few cases the existing network architectures have been utilised for limited-scale friendly user trials.

As said, the speed of innovation in the industry has driven early deployments into a set of proprietary or 'pre-standard' networking solutions. Virtually every femtocell vendor has developed its own networking technology, creating such a plethora of solutions that multi-vendor interworking with these early products is mostly not possible. For example, in early 2008 the Femto Forum catalogued around 15 distinct architecture combinations for GSM, UMTS and CDMA alone.

Even in this scenario of extreme architecture differentiation, a set of common best practice principles was adopted across different solutions, creating a common ground for initiating the standardisation work, particularly within 3GPP. In this field, the Femto Forum played a pivotal role in involving all the main industry players by providing a discussion forum where the basis of the femtocell networking standardisation has been defined (see Chapter 8), with the result that the first femto networking standards were drafted and agreed in just over a year, at least for UMTS where the new Iuh interface standard was published in December 2008 and further evolutions are in progress for CDMA, UMTS in an IMS variant, WiMAX and LTE.

5.3 Requirements

In order to address the challenges just described, the femto architecture must support as a minimum the following requirements:

* Scalability to millions of femtocell units in the same network.
* Operation over shared IP broadband links, like commercial xDSL, cable and FTTx with variable performance and contention with other home/office devices and data services.
* Hardened security of the femtocell units and of the architecture as a whole.
* Open-standard management interface, with reduced interactions with the femtocell management systems.
* Minimise additional load on the legacy operator infrastructure.

Deployment of femtocells can use shared IP broadband links owned by the network operator (*integrated* model) or by a different operator (*opportunistic* model). In the latter case, the network operator may not have control of the Quality of Service (QoS) provided by the IP broadband link and the femtocells must be able to dynamically react to such QoS changes. Many mobile network operators plan to use both the integrated and the opportunistic deployment models in the same network, or hybrid models where they have a collaborative approach with a set of IP broadband providers.

Most mobile operators plan to offer their femtocell services as a seamless extension of their macro network services, so the solution must support (as a minimum) full feature parity with the macro network, increased data capacity, similar QoS experience (call quality, voice quality and conversational quality) and full mobility between the femto and macro layers.[1]

As femtocells are introduced as an extension of the existing mobile networks, they must work with any existing legacy end-user terminal without the need for additional clients.[2] This means that the femto architecture must provide solutions for end-user terminal issues such as mobility to/from femto or femto zone identification that exclusively make use of the existing functionality[3] of the legacy end-user terminal.

Similarly, femtocells must not impact or degrade macro network quality or the services received by non-femto end-users.

---

[1] Branding reasons (keeping the brand promises) is behind the requirement to preserve all existing services and offer service continuity with the macro network. In most cases, the mobile operators have put a higher priority on full idle mode mobility and femto-to-macro network handover, leaving the macro-to-femto network handover as lower priority.

[2] For evolved femto services, like home integration, it is possible to introduce thin clients as presentation layers, but any service which is also available in the macro network must be available on the femto network without any terminal changes

[3] Some of the standards bodies, such as 3GPP for the UMTS/LTE Home NodeB, define improvements in the end-user terminal functionality, which provide a better support for femto services. While femtocell deployments will benefit from the introduction of such functionality, they must still work with any legacy terminal of the same technology, hence requiring solutions that work without end-user terminal changes.

Operators also want to be able to offer new home-zone differentiated services, beyond the pure tariff differential, such as home network integration, direct Internet access, VoIP/IP-PBX integration and femto-specific applications. The femto architecture must allow for such types of services offering local connectivity.

In general, the femto architecture must be fully standardised in order to enable multi-vendor networks, at least for the main functional elements:

- femto access point (FAP);
- femto gateway (FGW);
- femto management system (FMS).

Refer to Chapter 8 for further details on the femto standardisation progress.

Femtocells must also minimise the impact on existing macro core network elements, to avoid extra investment in what is usually considered a legacy network technology, providing fully standard interfaces to the core network elements and requesting no new features. In brief, the impacts on the mobile core network must be limited to capacity support and configuration, with efforts to minimise both.

Femtocells must comply with existing regulatory rules, which typically include emergency call support, end-user location determination, lawful intercept, packet inspection and in some cases the guarantee of uninterrupted service (within reason) as discussed in Chapter 9.

Finally, the femto network solution must have minimum impact on existing operational processes, from system integration to end-user provisioning, network management, billing, customer care and end-user support.

5.4 Femto Architectures and Interfaces

Broadly speaking, the femtocells and all of the network elements that provide the femtocell control, security, network integration and management functions compose the *femto network*. The femto network is an extension of the MNO macro network that provides local femto services to the end-user terminal.

The Femto Forum created a reference architecture including all of the elements of the femto network and their interfaces, shown in Figure 5.1. This diagram is generic to all of the major mobile networking technologies and serves as a means to compare alternative approaches on a common basis. It will be used in the remainder of this chapter to present the common architecture principles valid for any mobile networking technology.

The main functional elements of the femto network are:

- The *femtocell access point* (FAP), which is installed in remote premises, supports the radio interface to the mobile device and connects over the broadband backhaul to the MNO core network (CN). The FAP can be a stand-alone device connected to the existing home gateway providing the broadband IP connectivity or it can be integrated inside the home gateway itself.
- The *femto gateway* (FGW), which provides femtocell security, control, aggregation and standard interfacing with the MNO CN elements. The FGW hides the complexity and dimension of the femto network from the legacy macro network elements and a single FGW typically supports from tens of thousands to hundreds of thousands of FAPs. Usually the

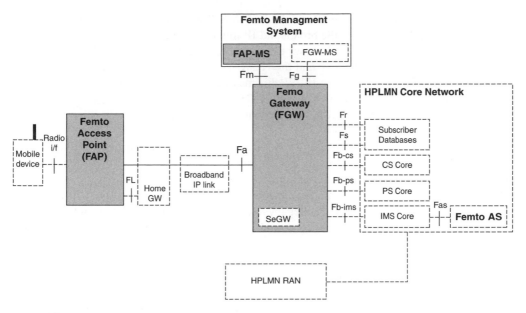

Figure 5.1 Femtocell reference model. *Reproduced by permission of Femto Forum Ltd*

FGW is composed of multiple network elements, with newly defined femto control/signalling elements and reuses where possible existing commercial elements such as:
- *security gateway* (ScGW) for the authentication and secure connectivity of the remote FAPs;
- *media gateway* (MGW) for the handling of the circuit-switched user plane over the IP layers;
- *authentication authorisation and accounting* function (AAA) to support FAP AAA procedures.

For the IMS-based architectures (see next section) part of the FGW functions are supported inside a *femto application server* (femto AS).

- The *femto management system* (FMS), which provides the management of both the FAPs in the remote environment over a standard interface and the FGW. The FMS is split into two elements:
  - *FAP management system* (FAP-MS), which manages the FAP elements and typically has a scalability in millions of units and must work on multi-vendor FAPs, hence the requirement for this element to be standardised;
  - *FGW management system* (FGW-MS), which manages the FGW in all its components and is typically not standardised.

In the Femto Forum reference model, these are the new interfaces introduced:

- The *Fa* interface between the FAP and the FGW over the broadband IP link.

- The *Fm* interface between the FAP and the FAP-MS. The Fm interface can connect directly to the FAP and FAP-MS over the broadband IP link or make use of the Fa secure layers to connect via the FGW. Figure 5.1 shows the latter implementation option only.
- The *Fl* interface between the FAP and the HGW in the remote premises. Where the FAP is integrated as a function inside the HGW, this may be an internal interface.
- The *Fg* interface between the FGW and the FGW-MS.
- The *Fas* interface between the IMS core network and the femto AS.

As the femto industry moves to multi-vendor networks, with the objective of having any vendor FAPs working with any vendor FGWs, the first targets for standardisation are the Fa and Fm interfaces and in many technologies also the Fas interface.

All other interfaces in the Femto Forum reference model are existing standard interfaces between the femto network and the legacy mobile core network and the mobile device, for which the Femto Forum reference model introduces new names to refer to them in a technology-independent way:

- The *radio i/f* between the mobile device and the FAP, e.g. in GSM solutions this is the Um interface.
- The *Fb-cs* interface between the FGW and the CS core, e.g. in RAN-based UMTS solutions this is the Iu-cs interface.
- The *Fb-ps* interface between the FGW and the PS core, e.g. in RAN-based UMTS solutions this is the Iu-ps interface.
- The *Fb-ims* interface between the FGW and the IMS core, e.g. in CDMA solutions this is the Wm interface.

5.5 Key Architectural Choices

A number of different architectures can fulfil the requirements described in the previous paragraphs, each one with specific tradeoffs. The three main architectural choices are about:

1. How to connect remote femtocells over the shared IP links to the MNO core network, which determines the femtocell network discovery and registration, security and QoS management.
2. Which layer of the core network hierarchy to connect to, which determines how much of the macro network functionality is delegated to the femto network, i.e. how 'flat' the femto architecture is.
3. How the functional split between the FAP and the FGW is best done, which also drives the choice for the FAP-FGW interface.

5.5.1 Connecting Remote Femtocells

Femtocells are a secure extension of the mobile operator's macro network into remote premises, typically inside end-user homes or offices or in other uncontrolled environments, and they connect over a shared IP broadband link such as provided by xDSL, cable or FTTx to the Internet and to the mobile operator's core network. As such femtocells must rely on hardened unit security and on secure and scalable procedures to connect to the core network secure gateway, as discussed in Chapter 7. The networking procedures described in this sub-section are only performed after the FAP unit has successfully completed the internal security checks at bootstrap.

5.5.1.1 Discovery and Registration

Femtocell deployments are typically 'unplanned', i.e. end-users are authorised to self-install the femtocell units at the agreed premises,[4] over any IP network and at any time after purchase. In some cases, e.g. when 'premium' services are offered to an enterprise customer, it may be the operator's personnel installing such units, but the procedures remain the same. When the potential scale of femtocell deployments is also considered, it is clear that femtocells must be able to autonomously find the best way to connect to the macro network.

This is best achieved in two steps:

1. A *discovery* step, where the FAP identifies the FGW to connect to.
2. A *registration* step, where the FAP and FGW are mutually authenticated and FAP services are authorised.

The *discovery* procedure usually includes using a hard-coded URL, specific to the mobile network operator, for a DNS query that defines the closest or best FGW to connect to. Typically this is the URL of an SeGW element of the closest or best FGW. The DNS query will then return the public IP address of such an SeGW. IP load-balancing techniques can be applied at this stage, but usually the FAP is connected to a first FGW, which then retrieves the FAP profile and may reassign the FAP to another FGW if necessary for efficiency or locality reasons.

The *registration* procedure provides the means for mutual authentication between the FAP and the FGW and authorisation of services including location authorisation. As explained in Chapter 7, in most implementations the security procedure is based on IKEv2 protocols (see reference (64)), with the use of PKI certificates (see reference (65)) for platform authentication and with the optional addition of a host security (usually SIM/USIM based). During the registration procedure, the FAP and the FGW also exchange information regarding the end-user current context such as location information, broadband IP network information, optionally augmented with performance and QoS management information.

5.5.1.2 Transport

Once discovery and registration procedures are successfully completed, the FAP and the FGW set up one or more secure transport layers for all subsequent control plane and user plane exchanges. Optionally, such layers can also transport the FAP management traffic, even though in some implementations the FAP to FMS interface is kept independent.[5]

---

[4] Some operators even enable the end-user to install units anywhere they like as long as this is inside the spectrum licensing area of such operator, or simply enable multiple locations for the same unit, e.g. a user can bring their femtocell to the holiday home (in the same country). Any femtocell suitable for this kind of deployment must be able to report its current location to fulfil regulatory requirements.

[5] This is a possible choice for mobile operators who also own the broadband IP network or have a fully trusted relationship with the broadband provider.

The typical choice is for IPsec with ESP (see references (66), (67)) and with UDP encapsulation to make it more robust to home gateway NATing. IPsec is universally considered to provide strong security and with the ESP tunnelling option provides a mean of going through any broadband IP networks. The price to pay is a performance reduction, as IPsec overheads are significant, especially for voice packets. Please refer to Chapter **7** for further information on the security architecture and to Section 5.6.1 for further information on QoS management.

5.5.2 Integrating the Femto Network with the Macro Network

Finding the right FGW to connect to, registering the FAP and setting up a secure transport layer constitute only the first step for the femto network to have access to macro network services.

The requirements placed on the femto architecture demand that no changes are needed in the legacy macro network elements, apart from capacity and configuration. This implies that interfaces between the femto network and the macro network must be based on existing standards. In brief the femto network must support *in toto* the functionality of a section of the macro network.

The second architectural choice thus is whether the femto network integrates with the macro network:

- as a radio access network element (*RAN-based architecture*);
- *OR* as a core network element (*CN-based architecture*).

While both solutions can fulfil the femto requirements as described in the previous paragraphs, this choice is driven by a number of factors and trade-offs, the most important being the choice between easier integration and immediate functional parity on the one side and level of innovation and cost-effective scalability on the other side. Operators who have given priority to the former have typically gone for a RAN-based solution, while operators who preferred the latter have gone for a CN-based solution, typically with an evolution of the IMS technology.

5.5.2.1 RAN-based Architecture

In a RAN-based solution, the femto network integrates with the macro CN as a RAN element, e.g. in UMTS this would mean that the femtocell network is seen by the UMTS CN as an additional RNC or set of RNCs (see reference (68) for details on the UMTS network architecture).

The RAN to CN interface is open and standard in all the main mobile networking technologies and it is typical for an operator to have multi-vendor integration over this interface. This makes the RAN to CN interface a natural candidate for femtocell integration.

The main advantages of the RAN-based architecture are:

- **Simpler integration**. The openness of the RAN to CN interface and the possibility to reuse the existing multi-vendor integration processes provide an excellent starting point for a fast integration process.

- **Immediate feature parity with the macro network**. When the femtocells are integrated as a RAN element, the upper signalling layers, the ones providing the end-user telecom services, go transparently between the end-user terminal and the CN elements, thus providing the same services set to the end-user. Additionally, this type of architecture can reuse the existing inter-RAN element mobility procedures to support full macro-to-femtocell network mobility.

The price to pay is in terms of making an investment in a new technology that still loads up the legacy networks. While this is seen as acceptable during early deployments, large-scale mature deployments may eventually create an issue especially if the operator is moving to next-generation networks in the meantime, i.e. all-IP-based.

In UMTS, the RAN-based architecture has been chosen for the initial standardisation work, with the emerging *Iuh* standards defining a world first for femtocells (see references (69), (70), (71)).

5.5.2.2 CN-based Architecture and IMS

In a CN-based solution, the femto network integrates with the macro CN as CN elements, e.g. in UMTS this would mean that the femto network is seen by the UMTS CN as an additional MSC and SGSN/GGSN. Inter-CN element interfaces are fully standard and open to multi-vendor (and multi-network) integration in all the main mobile networking technologies.

These CN to CN interfaces are typically more complex to integrate than the RAN to CN interfaces, but if the femtocell solution is based on the IMS network (see reference (72)), it provides the cost efficiency and scalability advantages typical of the all-IP networks. IMS networks are quite complex to integrate, so usually only operators who have already deployed commercial IMS services entertain using this architecture for femto networking.[6]

It is worth noting that all femto functions can be concentrated in new functional elements, so allowing for reuse of the existing Release 5 IMS networks unchanged (see reference (72)). In brief, the femto networking functions work as an additional IMS service with its own applications servers over a standard IMS network. A similar approach is possible for SoftSwitch (SIP/VoIP) networks.

The main advantages of a CN-based architecture using IMS are:

- **Cost-effective scalability**. The IMS networks elements are designed to make full use of IP networks and to have scalability in the millions on a single network element.
- **Maximum reuse/exploitation of the all-IP investment**, either IMS or VoIP. Not only does the femto integration make use of often underutilised IMS networks (or SIP/VoIP in some cases), but it provides a more efficient architecture to deliver services connected to other NGN users: the more end-users are migrated to NGN, the more benefits the operator reaps.
- **Access to IMS services from legacy end-user terminals**. As the IMS or SIP/VoIP client is supported in the femtocell, transparently to the legacy end-user terminal, IMS services can be extended (with limitations) to legacy devices maximising the customer base for such services from the start.

---

[6] The exception here is CDMA, as we will see later on in this chapter, where architectural constraints on the existing CDMA networks have dictated not using a RAN-based approach.

The price to pay is complexity of integration of the newly defined application servers with the legacy core network, as these elements are in all effects new CN elements to manage.

The Femto Forum has an initiative to define the long-term IMS femto networking approach, which for UMTS and CDMA will utilise 3GPP Release 8 new functions, such as the Multimedia Telephony (MMTel) and IMS Centralized Services (ICS) in order to provide seamless services (see references (72), (73)). It is worth noting that in order to implement such an architecture the whole IMS network needs to evolve to 3GPP Release 8 version,[7] thus making this architecture impractical for implementation for a few years to come, but an important step for those operators considering IMS as part of their long-term CN investment strategy.

5.5.3 Functional Split between the FAP and the FGW

The third architectural choice is to define how to split the RAN or CN functions supported in the femto network between the FAP and the FGW. In principle, all radio-related functions must reside in the FAP and all the macro network interfacing functions must reside in the FGW, with the remaining functions potentially residing in either of the elements. This also drives the choice of which type of interface to use between the FAP and the FGW.

Some of the early femto solutions have kept the FAP simple and moved all functions but radio to the FGW, very similar to picocell solutions, but most femto solutions have distributed much more functionality to the FAP and kept the FGW simpler, i.e. they apply a 'flatter' architecture. In particular, the requirements to have a FAP as self-configuring as possible and to provide support for new local services and local network access have driven the choice to flatter architectures, which in turn is enabled by the evolution to higher performing femtocell platforms. Flatter architectures also allow for more cost-effective scalability, which is important for mature femtocell deployments.

For example, in a UMTS RAN-based architecture, broadly speaking the choice would be between:

- A traditional, picocell-like approach where the femtocell provides the functions of a NodeB and the FGW of an RNC.
- A flatter approach, where the FAP provides the functions of a NodeB as well as the 'front-end' RNC functions, i.e. the ones that interact with the end-user terminal like radio resource management, ciphering and mobility support; while the FGW supports the 'back-end' RNC function interfacing with the core network as well as the necessary signalling aggregation. This is sometimes referred to as a *split RNC* architecture.

As already mentioned, in UMTS the industry consensus has focused on the latter option, which is now the basis for the Iuh standard. The detailed functional split applied for the Iuh standard can be found in (69).

For CN-based architectures, the most common choice for the functional allocation has been for a *split CN element* case, where the 'front-end' CN functions interacting with the end-user terminal are supported in the FAP, and the CN interfacing with the macro network is supported

---

[7] Current IMS deployments are typically based on 3GPP Release 5 with few networks already implementing selective functional elements of 3GPP Release 6 and 7.

in the FGW. This has been the industry choice for CDMA technology and it has also been adopted in some UMTS deployments.

5.5.3.1 FAP-FGW Interface

The FAP-FGW functional split also defines the type of interface between these elements. The lower layers, IP layer and below, have been addressed previously in the Transport subsection 5.5.1.2. Now looking at the upper layers' control and user planes we have the following main cases:

- For RAN-based solutions, where the FAP is kept to a minimal functional set, the FAP-FGW interface is derived from the one used between the macro network cells and RAN controller, e.g. the Iub interface in UMTS, with extensions to make it work over a shared IP link.
- For RAN-based solutions that adopt a flatter architecture, the FAP-FGW interface is derived from the one between the macro network RAN and CN functions, e.g. the Iu interface in UMTS again with extensions to make it work over a shared IP link, as captured in the Iuh standards.
- For CN-based architectures, the FAP-FGW interface, rather than being derived from the CN to CN element interfaces,[8] has typically been redefined to use IMS procedures over the Wm interface (see reference (72)). In this case the FAP interworks with the 'front-end' CN procedures with the IMS procedures on the FAP-FGW interface. Similarly the FGW interworks with the 'back-end' CN procedures.

5.6 Other Important Femto Solution Aspects

Following the key architectural choices, there are a number of important aspects that the femto solution must address:

- Ensuring proper end-to-end security of femto systems, macro network and the end-users, as addressed in Chapter 7.
- Providing remote femtocell management, as addressed in Chapter 6.
- Supporting end-to-end QoS.
- Providing local access to home networking, direct access to the Internet and femtocell integration with VoIP networks and IP-PBX.
- Providing support for femtozone services, with billing differentiation and home-zone indication: here introduced and further explained in Chapter 11.
- Supporting full mobility with the macro network and between femtocells.
- Providing femtocell location retrieval/locking and any other regulatory requirements: here introduced and further explained in Chapter 9.
- Supporting enterprise and open spaces offers.

---

[8] An early femto solution that was adopting a set of the CN interfaces for the FAP-FGW interface has been withdrawn.

5.6.1 End-to-End Quality of Service

Femtocell deployments make use of shared broadband IP links to connect the FAP in remote locations. Such links introduce a number of QoS variables and have lower performance compared to the typical macro network links. Additionally, when the mobile network operator deploys femtocells in an opportunistic model, i.e. reusing an existing broadband IP link offered by a different operator, there is little coordination of QoS possible on the IP links.

Most femtocell technologies provide good quality voice calls and sufficient support to data services when the broadband IP link provides a minimum performance of:

- less than 150 ms round-trip delay;
- less than 50 ms jitter;
- less than 1% packet loss;
- at least 1 Mbps in the downlink direction, i.e. from the broadband IP provider network to the HGW;
- at least 256 kbps in the uplink direction, i.e. from the HGW to the broadband IP provider network.

In general, the performance offered by the typical broadband IP link matches the requirements defined above, thus it is sufficient to support good quality femtocell connections, but at times of congestion it may introduce noticeable degradation. Furthermore, some older broadband IP links may have limited bandwidth available in the uplink direction. In these conditions, the femtocell technology must introduce countermeasures to mitigate the performance impacts to the maximum possible degree.

Most femtocell solutions adopt VoIP techniques to provide consistent voice call quality, like using variable buffer sizes to recover link jitter in each direction, and are robust to delay and to packet loss beyond the values shown above.[9]

The majority of femtocell solutions have also introduced some form of QoS management techniques. As a minimum these consist of tagging each IP packet with its QoS class (see reference (74)) and prioritising these packets inside the FAP and FGW queues and in some cases reducing the QoS for lower priority services, like best-effort data sessions. In any case, in opportunistic deployment models, but frequently also in integrated deployment models, the QoS tagging is ignored inside the broadband IP link and the HGW. In order to offer premium services, mobile operators are thus progressing to reach agreements with the broadband IP providers to support QoS end-to-end on the shared IP links as well as control of IPsec traffic policy management.

The operator community has put the requirement to support up to four voice calls over a 200 kbps link, where typically the bandwidth limitation is the uplink (FAP to FGW) direction. To achieve this, multiplexing technologies between the FAP and the FGW (specifically the MGW function of the FGW) are necessary, as with the IPsec overheads a single voice call

---

[9] Where the broadband IP link introduces additional round-trip delays of more than 100 ms, the mean opinion score of the voice call quality starts being affected in a similar way to the one experienced in some intercontinental voice calls. The voice call can be maintained also for values of 500 ms and higher, but the parties in the call must adapt their behaviour, i.e. leave some time before replying.

Packet loss of more than 1% typically is perceived as voice quality degradation ('metallic voice') and it is difficult to maintain a voice call when packet loss exceeds 5% in either way.

would need a peak rate of between 63–84 kbps depending on the lower layer technology.[10] Multiplexing can work by putting multiple voice frames, one from each one of the calls in parallel, on top of the same IPsec outer packet, much reducing overheads at the expense of managing some additional complexity.

5.6.2 Local Access (Data and Voice)

Femtocells already adopt a 'flat' network architecture where many network functions are performed at the edges of the network, i.e. in the FAP. For network operators seeking to maximise the benefits of the flat architecture, the next logical step is to require that the FAP manages the *local access* connections to other devices, to the Internet and potentially also to other types of networks such as PSTN and VoIP. The Femto Forum adopts the name of *local IP access* for the data local access. Because of the efficiency gains in unloading the mobile operator core network from this type of traffic, these technologies are also referred to as *data offload* and *voice offload*.[11]

As femtocells use licensed spectrum technology, the mobile operator must retain control of the local access authorisation, rule and policies and must be able to deactivate it in selected cases (see also Chapter 9 for the regulatory implications).

5.6.2.1 Local IP Access

The local IP access encompasses the direct connection of mobile device data sessions to other data devices in the home/office and the direct connection to the Internet or to other IP services in the IP cloud *without* going through the FGW or the mobile operator core network.

The local IP access increases the data session's performance, by cutting away delays and bottlenecks generated by the core network, and creates new service opportunities for integrating mobile devices on femtocells to the rest of the home or office network.[12]

In order to provide local IP access, an FAP must support:

- Radio ciphering, as only a decrypted user data plane can be offloaded. This means that the radio encryption keys must be delivered to the FAP by the FGW during the mobile device authentication and mobility procedures.
- IP routing capabilities for the data offload. Here the mobile operator has a choice of whether to configure dedicated data sessions for the local IP access, or for more seamless services, to apply IP routing masks[13] for selective data offload during sessions that can contain data for both the mobile core network and the local devices/Internet.

---

[10] This bandwidth is the peak rate calculated for a 20 ms sampling voice call, ignoring any silence suppression which kicks in during one-way silence periods of the call, and for a DSL technology, taking into account the ATM framing and where the DSL transport may be either PPPoA or PPPoE, with the latter requesting the highest bandwidth.

[11] It is estimated that at least 90% of the data traffic generated by mobile devices is terminated in the Internet, thus adding a massive load on mobile operator networks for accessing services outside these networks. For unlimited data plans, this traffic adds a lot of cost and no revenues, thus the eagerness of mobile operators to offload it locally.

[12] Connecting mobile devices inside the local LAN is particularly important for the enterprise case, as typically nearly half of the data traffic generated by business users is for direct connections to the enterprise servers.

[13] For example an operator may configure a mask with all public IP addresses targeted for local data offload as well as a configurable subset of the local LAN addresses.

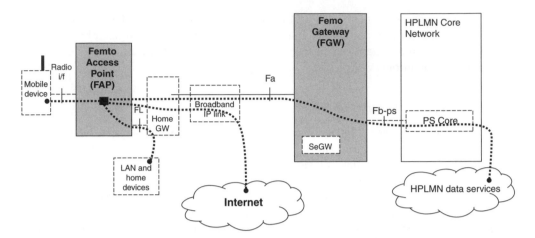

Figure 5.2   Local IP access possible data connectivity

- Firewalling, IP filtering and packet inspection for security reasons and to implement operator policies, e.g. parental control.

Local IP access creates potential security issues, as it exposes an additional interface inside the home. Please refer to Chapter 7 for further details on the security implications. Members of the Femto Forum have been examining standardisation proposals for local IP access.

Figure 5.2 shows the local IP access routing capabilities for a mobile device connected to a FAP. The data path to the core network goes over the Fa interface secure layers, while the LAN data path and the direct Internet data path are unencrypted.

5.6.2.2   Voice Offload

A similar concept to local IP access can be applied to voice calls, and in general to any circuit-switched call. An important difference here is that there are many variants of voice networks, and so far voice offload has been announced for connectivity to the PSTN, to existing VoIP fixed-line networks and for PBX integration inside an enterprise.

Voice offload creates the opportunity for local calls, important especially in the enterprise case, and for a true fixed-mobile convergence and integration with the fixed-line voice services. For example, a mobile device camped on a femtocell could be reached both via its mobile number (and the mobile core network) and via a VoIP fixed-line number, thus the mobile device can double up as a cordless phone inside the home/office. Similarly, the same mobile device can place calls via the mobile network or to the VoIP network, e.g. by means of using a prefix to the phone number. Some operators are targeting to offer a zero-cost Friends & Family service by using automatic voice offload to a selected set of fixed-line numbers.

In order to provide voice offload, an FAP must support:

- Radio ciphering, as already seen for the local IP access, before termination of the upper layer call set-up signalling, in order to detect called party and calling party numbers.

- Voice routing. Routing of mobile originated calls can be set as automatic, e.g. to selected VoIP fixed-line numbers, or by using a prefix to select the VoIP network. In enterprise cases, with PBX integration, calls to local extensions are routed via the voice offload.
- A voice client to interconnect to the voice network of choice. Typically this is a VoIP or SIP client for connection to VoIP networks, IMS networks or IP-PBXs, but it can also include support for a PSTN connection.[14]

5.6.3 Femtozone Services

While connection via a femtocell can be made completely seamless to the end-user, most mobile operators plan to offer special *femtozone* services and tariffs in order to entice the end-users to adopt femtocells, increase traffic, offer new services and ultimately to increase ARPU and loyalty.

Typically mobile operators offer reduced tariffs for end-users connected to a femtocell, e.g. charging fixed-line tariffs for mobile originated calls or offering all-you-can-eat data plans while in the home.

The higher performance and very limited cost of supporting high-speed/high-volume data services on a femtocell, especially when local IP access is used,[15] combined with reduced or zero charges for the end-user, make possible a number of services that currently exist on the Internet but have limited support on mobile devices, such as video streaming and sharing. Mobile operators anticipate that the end-users will also learn to use a selection of these services while on the macro network where higher charges may apply.

First of all, the mobile operator must be able to notify the mobile device of when it is camping on a femtocell. The end-user notification must work on any legacy devices[16] and typically involves showing a different network identifier on the screen (like 'MobileOperator@home'), which can be supplemented by welcome messages when the mobile device registers on a femtocell and optionally with audio notifications and audio messages at the beginning of voice and video calls, or by means of SMS notification for any new packet data session. Note that where a different tariff is applied, it can be a legal requirement to show unequivocal identification of the applied tariff.

By using the local access functionality, mobile operators are planning to offer a set of new services that connect mobile devices to the home network. Some examples of these services are:

- Enabling mobile devices to synchronise with the home entertainment system, to store favourite programmes locally for later playback when the user is out of the home, e.g. travelling.
- Connecting motion triggered security cameras or (video) baby monitors to a set of mobile devices registered on the femtocells.

---

[14] In this case, typically the HGW supports a PSTN connector and a software adapter to interwork with the FAP.

[15] For data traffic that is routed directly to the Internet by using local IP access functionality, virtually the cost to the operator is the same as for fixed-line broadband IP traffic. In the case of the opportunistic deployment model, that cost is zero to the mobile operator.

[16] The latest mobile devices, such as Release 8 mobile phones in UMTS, offer new capabilities that make it easier to notify and display the femtozone identifier to the end-user.

Because of all the possibilities that a femtocell offers to integrate the mobile device with any of the existing triple-play services (fixed-line voice, broadband data, IPTV), femtocells can be an enabler for true *quadruple-play* services. Please refer to Chapter 11 for further details.

5.6.4 Mobility

Mobility to/from the macro network is a necessary requirement in order to provide seamless femto services to the end-users. Due to the potential scale of femtocell deployments and the limitations of existing mobile devices, which are able to identify only a few neighbour cell characteristics, mobility in femtocells presents difficult challenges.

5.6.4.1 Idle Mobility

For an end-user terminal in idle mode, mobility is driven by the measurements of the neighbour cells' radio signal quality and strength. The end-user terminal takes autonomous decisions on when to apply an idle mobility procedure. The key is to provide a sufficient neighbour list to allow macro-to-femto, femto-to-macro and femto-to-femto idle mobility.

In the *macro-to-femto network direction*, it is necessary that all radio characteristics used by femtocells are advertised in the macro cell neighbour lists as potential neighbours. This means basically that the femtocells must use a limited range of radio characteristics, typically fewer than 10 combinations, such that there is enough space in macro cells' neighbour lists to list them all.[17]

Once the end-user terminal has performed the idle mobility procedure, it will typically register in the femto network in order to access femto services. This is done, for example in GSM and UMTS, by means of a Location Update procedure (see reference (75)).

In the *femto-to-macro network direction*, two approaches are possible:

- Auto-configured neighbour list, where the FAP must be able to measure the macro network signal in the local environment in order to reconstruct a full neighbour list. Optionally the FAP can also choose independently most of the radio parameters, following the rules or policies defined by the MNO and reporting such choices to the FAP-MS.
- Centrally configured neighbour list, where the neighbour list is provided to the FAP via the FMS and a central application that calculates the list based on the presumed FAP location. Optionally the FAP can directly provide the location information, e.g. by means of a GPS receiver, or by other means as described in Section 5.6.5.

The choice above is left to implementation preference, but as the latter is heavier on operational processes and prone to failures due to the dependency on the location information, most femtocell vendors have adopted to have some form of auto-configured neighbour lists. Once the femto neighbour list is defined, the FAP broadcasts it on the radio channels as usual in order for the end-user terminal to be ready to apply idle mobility.

---

[17] In some cases, as in large indoor coverage solutions, it is possible to have only some of the femtocells seen as neighbours by the macro cells and have the remaining ones as unseen, i.e. this second set can only be reached in idle mobility after the end-user terminal has relocated to a femtocell of the first set.

In the femto-to-femto case, there is a similar choice between auto-configured and centrally configured neighbour lists, and in the simplest case it is sufficient to augment the macro neighbour list with the list of all the other femto radio characteristics combinations. Clearly, in the auto-configuration case this femto neighbour list can be accurately measured and limited to only the ones seen.

5.6.4.2 Handover

The femto handover support presents some additional challenges compared to idle mobility, due to the fact that during a handover the network must know the identity of the target cell unambiguously in order to prepare it for the incoming handover.

In the femto-to-macro handover case, this can be easily done by extending the neighbour list to include not only the radio characteristics of the neighbouring macro cells, but also their full identity. In the auto-configuration case, it is the FAP itself that detects the macro cell identity from the macro cell broadcast channels and then uses it during the handover procedure. Clearly, any local access service is excluded from continuity during a handover with the macro network.

In the macro-to-femto handover case, it is impossible for the macro cell to know unambiguously the identity of the target femtocell. This is due to the fact that potentially there can be up to hundreds of femtocells in the area covered by a single macro cell,[18] in which case it is impossible to differentiate them merely according to their radio characteristics as reported by the end-user terminals.[19] Furthermore, if femtocells were to be known by identity in the macro cells, this would equate to an extremely heavy and dynamic configuration effort.

In general mobile network operators have given higher priority to femto-to-macro handovers, as this is a necessary procedure for service continuity, while in the macro-to-femto case the macro network signal may still be good enough to provide service indoors.[20] However, this is only a temporary trade-off that the most innovative operators are willing to accept in order to speed up their femtocell deployments, but all require support of such functionality in the longer term.

So the debate is still on regarding how best to approach the macro-to-femto handovers. In general it is accepted that this type of handover will be *ambiguous*, i.e. multiple femtocell targets must be prepared to receive an incoming handover and there is a chance that, when the mobile device tries to connect to one of them, the femtocell is at capacity and rejects the incoming handover. This also means that the FGW must perform filtering of the handover requests to avoid spurious attempts, e.g. by end-users who are not subscribed to a closed-access femto service or to access to the femto network in general, and to narrow down the number of target femtocells that must be prepared in parallel, e.g. by using information about the femtocell location or heuristics on previous attempts from the same mobile device. Once an ambiguous handover is performed, all of the FAPs that were prepared to receive the handover and did not receive it must release the reserved resources and clear the FGW-FAP context for that end-user.

---

[18] Typical macro networks for medium-large countries have deployed in the order of tens of thousands of macro cells, while a mature femtocell deployment in the same country may scale up to millions of units.

[19] The very latest mobile devices, such as Release 8 UMTS and LTE phones, are able to report additional characteristics including the cell-ID of the measured neighbours, thus making it possible to have unambiguous macro-to-femto handovers. Unfortunately, it will take a few years before these devices reach significant penetration.

[20] This in the case femtocells are deployed not as a coverage means, but for capacity and new services reasons.

In the femto-to-femto handover case, it is possible for the originating FAP to determine the full identity of the target FAP by means of auto-configuration, i.e. detecting the neighbour FAP identity from its broadcast channel information or by using some central configuration and information distribution via the FAP-MS. This type of handover is typically supported in most femtocell solutions.

5.6.5 Femtocell Location

Femtocell (FAP) location is needed for a number of reasons:

- Licensing reasons, to exclude that the FAP operates in areas where the mobile operator does not have spectrum, or simply to detect in which licensing area of the same mobile operator the femtocell is in order to select the operating frequency accordingly.
- Regulatory reasons, as in most countries where emergency calls must be provided with an indication of the end-user location.
- Location services possibilities, in order to provide location information to any services such as calling a taxi or finding the local street map and facilities.

Depending on the location precision needed, different solutions are possible. For detecting an in-country/out-of-country location and for the less demanding location services, it is sufficient to analyse the macro network signal detected by the FAP for triangulating the FAP position and possibly also check on the IP address used by the broadband IP link.[21]

Where regulatory and licensing reasons demand a high location precision, typically two solutions have been adopted:

- Relying on the broadband IP link line identifier, tying the FAP to work only on such a line and having access to the street address information of where that line is provided. This solution requires a tight integration of the mobile operator femto solution with the broadband IP offer and thus cannot be offered in an opportunistic deployment model yet, even though the broadband IP providers are progressively upgrading their networks to offer this location service as a paid-for service via an IP address interrogation.
- Adding a GPS receiver to the FAP to detect the FAP location with a very high precision. The main issue with this solution is that the GPS signal tends to be weak indoors and it does not work properly in a small, but significant, minority of cases, or needs a significant time before detecting the location after a power cycle of the FAP, or needs an external antenna, sometimes with a long lead. This is on top of adding extra cost and power consumption to the FAP unit.

As there is no single 'silver bullet' solution for femtocell location determination, the femto industry is still debating which one is the best and current deployments adopt a mix of the above techniques that varies country by country. See Section 10.3 for further discussion of the potential solutions.

---

[21] Typically broadband IP providers use country-specific IP address ranges that the FGW can check for authorisation during the FAP registration.

5.6.6 Enterprise and Open Spaces

Femtocells were initially targeted at consumer offers, but it was immediately clear that this technology presents a number of benefits for the enterprise case and for the coverage of open spaces.

Small office/home office (SOHO) business users can have immediate benefits by utilising a consumer unit, typically with local access enabled in order to connect to local LAN servers.

Small and medium enterprise (SME) and large enterprises need a different solution as multiple units are necessary to provide the necessary coverage and capacity.

Some femtocell vendors have developed higher capacity/higher power units, the class 2 devices defined in Section 1.4 and sometimes called *super-femtos*, which resemble picocells for the level of service provided, but connect to the macro network as femtocells. Compared to picocells these units present obvious advantages in that they do not need dedicated links, but they are more expensive than consumer femtocells, require a managed deployment and may be under regulatory site approval.

Other femtocell vendors have preferred to develop a distributed solution, with a self-organising grid of consumer femtocells (the class 1 devices defined in Section 1.4) that act as a bigger unit. In these solutions, femtocells can be deployed by the enterprise personnel autonomously, very much like a grid of Wi-Fi access points. The advantages of this solution are in reusing cheaper consumer units and much reducing the deployment and operational costs, but this can result in capacity limitations where a high number of users aggregates in the same location.

Deployments have been announced for both solutions. As the platform capacity evolution is increasing the number of calls and the data rates supported by consumer femtocells, the capacity gap between the two classes of device may reduce.

In both cases, femtocell solution must implement femto-to-femto handovers and femto-to-macro handovers. Given the importance of macro-to-femto handovers for business users (think of the case where there is no macro signal inside the building), in some cases the FAP units at the perimeter/entrance of the building are managed as part of the macro network radio management procedures, so that their unique identities and radio parameters are known in the macro network as neighbouring cells.

Some mobile operators prefer to offer enterprise solutions as open access, i.e. open to any subscriber of the mobile network, while others prefer to offer them as closed access, in order to add enterprise specific services such as LAN and PBX integration by using the local access functionality. Hybrid solutions are also possible, with enterprise users authorised for full services and guest users authorised for limited services and connected exclusively to the core network.[22]

Covering open spaces, such as railway stations, airports or shopping malls, with femtocells is very similar to the large enterprise case, but for open spaces only open access mode is ever used. Mobile operators are also thinking of special services for the open space case, e.g. by adding location information or creating captive portals that advertise local businesses, very much like when connecting to a Wi-Fi hotspot.

---

[22] For example, enterprise users can be authorised for maximum data rates, LAN access, direct Internet connectivity and PBX access, while guest users receive limited data rates and are not authorised for any local access service.

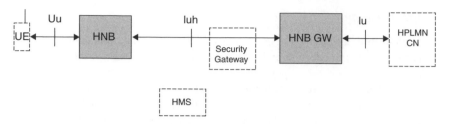

Figure 5.3   Iuh interface reference model. *Source:* 3GPP

5.7 UMTS Femtos

The first femto networking standards published are for UMTS femtocells, where the Femto Forum focused on many of the 2008 and early 2009 activities.

The 3GPP body approved the first version of the *Iuh* (Iu for Home NodeB) standards as part of 3GPP Release 8 during the December 2008 meetings and kept working on a number of refinements over the following months, see Chapter 8 for more details.

The Iuh reference model is reported in Figure 5.3.

Compared to the reference architecture defined in the Femto Forum and shown in Figure 5.1, the 3GPP terminology used in Figure 5.3 makes the following changes:

- the FAP is called *Home NodeB* (HNB);
- the FGW is called *HNB Gateway* (HNB GW);
- the Fa interface is called *Iuh* interface.

5.7.1 Iuh protocol stacks

The Iuh protocol stacks are derived from the Iu protocol interface with some simplifications and additions to make it specific to femto services. Figure 5.4 shows a reference diagram for all protocol stacks used in the Iuh solution.

The **HNB authentication** with the HNB GW uses IKEv2 procedures (see reference (64)) with optional dual authentication:

- A mandatory *platform* authentication, which makes use of PKI certificates (see reference (65)). This requires that the HNB is loaded with an individual certificate, signed by a recognised Certificate Authority (CA). The HNB GW uses a root certificate assigned by the same CA to perform a mutual authentication with the HNB.
- An optional *host* authentication, which makes use of a SIM/USIM card. This second authentication provides layered security to the mobile operator, i.e. the mobile operator can trust the vendor security processes for the platform security while keeping tight control and simpler operational integration with existing mobile device processes for the host security. The host authentication is performed as part of the same IKEv2 procedure with the platform authentication. Where this type of authentication is not used, the HNB GW makes use of the platform authentication to authenticate also the host services.

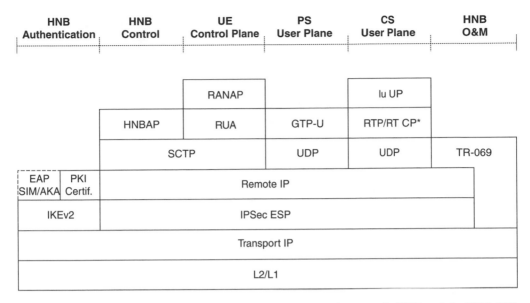

Figure 5.4 Iuh interface protocol stacks. All protocol stacks are between the HNB and the HNB GW, but the HNB O&M which is between the HNB and the HNB-MS. RTCP is optional in the figure

After successful mutual authentication, the HNB and HNB GW set up an IPsec secure transport layer, with two security associations: one for the uplink and the other for the downlink direction. All end-user traffic is consolidated over the same IPsec tunnel. Please refer to Chapter 7 for further details on authentication and transport security.

Compared to the Iu interface as defined in (76), the Iuh **control planes** introduce some simplifications by removing the M3UA and SCCP layers, as these two protocols were defined to connect large network nodes on high-performance links, and introduces:

- A RANAP User Adaptation (*RUA*) layer (see reference (70)), which provides transparent transport for RANAP messages and error handling. This layer was introduced in order to avoid any changes for the RANAP signalling, as RANAP signalling connects with the legacy MSC and SGSN.
- A femto-specific application protocol called Home NodeB Application Protocol (*HNBAP*) layer (see reference (71)), which provides support for the HNB registration procedures with the HNB GW and the UE registration with the HNB GW, this on top of error handling.

The SCTP layer makes use of the IPsec tunnels as transport.

The Iuh **user planes** upper layers are the same as the Iu interface over IP transport for both packet switched (PS) and circuit switched (CS), with the IP layer transported over the IPsec tunnels.

The **HNB O&M** uses the TR-069 management protocol (see reference (77)) between the HNB and the HNB-MS. The TR-069 layer is authenticated with a PKI certificate, typically the same one used for platform authentication, and can make use of the secure IPsec tunnels or connect directly over IP links. See Chapter 6 for more details.

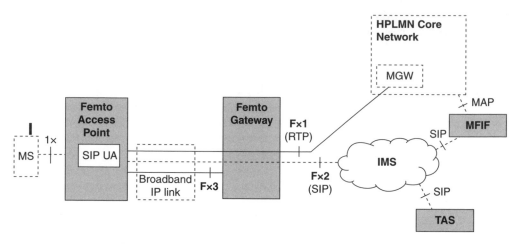

Figure 5.5   CDMA 1x femto reference model. *Source:* 3GPP2

5.8 CDMA

CDMA mobile operators have been the first to offer consumer femtocells for nationwide deployments. Most of the initial commercial deployments have been announced for the CDMA 1xRTT technology variant.

For CDMA deployments (including 1xRTT and EV-DO), the architecture choice has been for a CN-based solution that uses an IMS network to support femto services. The reference diagram for CDMA 1x is shown in Figure 5.5. CDMA 1x femto standards are being published by 3GPP2, see (78).

The FAP hosts a SIP user agent (SIP UA) to interwork the 1x procedures on the mobile device side with a SIP/RTP interface with the core network. The FAP terminates all 1xRTT signalling and user plane EVRC packets locally and converts them into SIP and VoIP traffic, respectively.

The FGW function as defined in the Femto Forum reference architecture (Figure 5.1) is subdivided into three elements:

- A *femto gateway*, which authenticates the FAP and manages the secure layers over the broadband IP link. In the CDMA 1x definition, the femto gateway is not involved in upper layer signalling procedures.
- A *telephony application server* (TAS), which provides the MSC-equivalent services for the 1xRTT mobile camping on an FAP. This element is implemented as an IMS application server and uses SIP interfaces.
- A *MAP-femto interworking function* (MFIF), which provides the signalling interworking between the FAP supporting the 1xRTT, the IMS environment (including the TAS) and the CDMA core network elements (1x MSC, HLR and SMSC).

The IMS cloud is intended to be a standard Release 5 IMS that supports the SIP signalling switching between all network elements. The media gateway function is in common with the CDMA 1x core network, as shown in Figure 5.5.

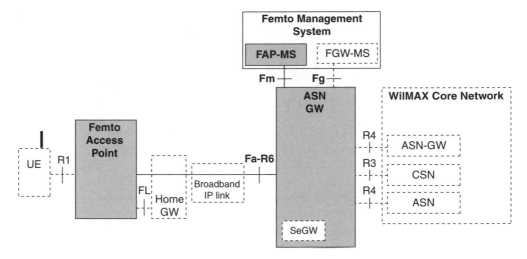

Figure 5.6   Proposed WiMAX femtocell architecture. *Reproduced by permission of Femto Forum Ltd*

The CDMA1x femto standards introduce three new interfaces:

- *Fx1* interface, which supports the RTP bearer carrying the end-user EVRC payloads between the FAP and the media gateway, via the femto gateway.
- *Fx2* interface, which carries the user SIP signalling between the FAP and the IMS core network, via the femto gateway.
- *Fx3* interface, which provides the IPsec tunnel between the FAP and the femto gateway as well as the FAP-FGW discovery and registration functions.

5.9 WiMAX

The WiMAX Forum, along with the IEEE for the underlying standards, is working on standards for WiMAX femtocells following a set of requirements from service providers (79). One possible reference architecture for WiMAX femtos is shown in Figure 5.6.

WiMAX is a data-only technology, so the WiMAX femto solution does not include any circuit-switched service interworking.[23] The WiMAX femto solution closely follows the existing WiMAX network architecture, with modified access service network gateway (ASN GW) also functioning as a femto gateway to terminate the secure IPsec layers with the FAP.

The AP-ASN interface (R6) is modified to introduce the femto-specific procedures for FAP-ASN GW discovery, registration and set-up of the IPsec layer. The FAP-ASN GW interface is named *Fa-R6*.

---

[23] Circuit-switched services over WiMAX are supported with mobile device clients that interwork directly with servers in the IP cloud, using the WiMAX network purely as an IP transport.

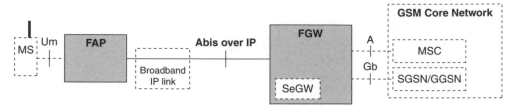

Figure 5.7   GSM 'Abis over IP' femtocell architecture

5.10 GSM

A number of femto GSM solutions are currently available on the market and are being deployed, but the GSM/UMTS mobile operators have put a priority on UMTS femtocells. This is due to a number of facts:

- GSM indoor penetration is already fairly good in most countries.
- GSM networks have enough spare voice capacity now that end-users are progressively migrated to UMTS.
- There are limitations to the type of data services that GSM can support, due to the lower bit rates available.[24]

In general, mobile operators prefer to invest in UMTS networks and focus on migrating GSM users to the new technology.

GSM femtocell solutions are mostly targeted on coverage enhancement of the macro network, for particularly difficult indoor situations. As a result, no significant GSM femto standard activity has happened yet.

Current GSM femtocell solutions are either derived from earlier picocell technologies, for higher capacity/higher power solutions, or from mobile device platforms, for very low capacity but more cost-effective small-cell solutions.

In general, all GSM femto solutions use variants of the 'Abis over IP' technology, shown in Figure 5.7. This is a RAN-based femto solution where the FAP plays the role of the basestation transceiver system (BTS) and the FGW of the base station controller (BSC), see reference (80) for details of the GSM architecture. The Abis interface, which is the interface between macro network BTS and BSC, is modified accordingly to function over the shared broadband IP link and transported over a secure IP layer. The Abis interface is left as a proprietary interface in the GSM macro networks. As a consequence, so far there have been no announcements of multi-vendor interoperability of GSM femtocells and gateways.

It is worth noting that in the GSM technology, the data user plane ciphering happens between the mobile device and the GSM core network elements (the SGSN specifically, see reference (81)), so the 'Abis over IP' GSM femto solutions have inherently no possibility of supporting local IP access for home/office and Internet direct data connectivity. An early femto solution, which also included the SGSN functionality in the FAP to solve the local IP access problem, has since been withdrawn from the market.

---

[24] Over a dedicated carrier frequency, GPRS can deliver close to 100 kbps max bit rate and EDGE can deliver around 300 kbps max bit rate, while typical broadband data services require at least 1 Mbps to work properly.

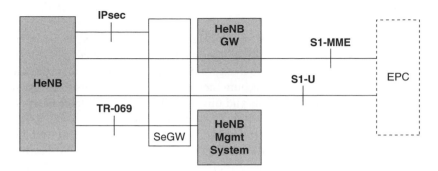

Figure 5.8   LTE femtocell architecture. *Reproduced by permission of Femto Forum Ltd*

5.11 LTE

LTE is gaining momentum in the mobile telecommunications industry, with a few trials announced that are evolving into commercial solutions. Compared to the other mobile network technologies, in LTE there is as yet no significant installed base, so the business proposition for femtocells is quite different, in that it is targeted to complement the early LTE deployments rather than targeting legacy devices for better services.

The LTE standards promise a common evolution path for a number of mobile network technologies, including UMTS, CDMA and WiMAX (see reference (82)). As LTE femtocells are in the very early stages of development, the femto industry is using this chance to prepare an LTE femto standard before any deployment. The discussions for the LTE femto standards are undergoing in the Femto Forum, in NGMN Alliance and in 3GPP. While the architecture has not been finalised, there is a strong consensus to keep it as flat as possible, following the principles of 'all-IP' networks adopted in the LTE standards. The debate is still going on as to whether there is a need for a signalling aggregation element or whether the evolved packet core (EPC) itself should be able to support femtocells directly. The reference LTE femto architecture is shown in Figure 5.8.

Compared to the reference architecture defined in the Femto Forum and shown in Figure 5.1, the 3GPP terminology used in Figure 5.8 makes the following changes:

- The FAP is called *Home evolved NodeB* (HeNB).
- The FGW is called *HeNB Gateway* (HeNB GW) and the Security Gateway (SeGW) function is separated.
- The FAP-MS is called *HeNB Management System* (HeNB MS).

The HeNB to SeGW lower layers are based on IPsec and IKEv2, with the same functions as described in the UMTS femto (Section 5.7). The HeNB management interface with the HeNB MS is based on TR-069.

The interfaces between the HeNB and the EPC are the standard S1-MME and S1-U, with the HeNB GW optionally providing aggregation function for the S1-MME. The S1-U interface adopts a direct tunnel approach, but optionally also this interface can be aggregated by the HeNB GW. In this case, the HeNB GW may also provide support for user plane multiplexing, for efficient transmissions over limited bandwidth links.

5.12 Conclusions

With femtocells' promised benefits of cheap coverage, capacity and new services opportunities becoming real over the last two years, the femto industry has not waited for standardisation to happen before the first commercial launches, resulting in a range of femto architectures. In spite of the huge variety, operators' requirements have driven vendors to define solutions that are fairly similar in functional terms and this natural convergence helped and is helping to drive the femto industry to a fast definition of common standards.

The femto network is organised into three main functional blocks: the FAP installed in the remote premises; the FGW providing the core network interface; and the FMS providing the FAP and FGW management. The industry preference has been for 'flat' architectures, with most functions supported in the remote FAP. The remote FAP automatically discovers the best FGW to connect to over the broadband IP link, registers with the FGW and FMS and sets up the secure layers.

The femto network solutions are easily integrated into existing mobile networks as they use pre-existing standard interfaces to connect to the mobile devices and to the legacy mobile core network. The architecture choice has typically been between a RAN-based solution, where the femto network integrates with the macro network as a RAN element; and a CN-based solution, usually associated with IMS, where the femto network integrates as another CN element. RAN-based solutions are preferred for simplicity of network integration, while CN-based (IMS) solutions are preferred for cost-effective scalability and all-IP synergies.

Femto architectures have addressed the scalability, security and connectivity over shared IP links, minimal impacts on existing operational processes and remote management challenges that are typical of mass-market deployments and enable both integrated and opportunistic deployment models, i.e. with or without agreements between the mobile operator and the broadband IP provider. Femto architectures have solved the difficult issues of femto-to-macro and femto-to-femto mobility, location retrieval/locking and support local access and new services in order to serve the consumer, enterprise and open spaces segments.

There is a strong drive in the femto industry to standardise all interfaces to the FAP. The Femto Forum fostered rapid early discussions between operators and vendors, which led to the definition in 3GPP of the Iuh standards for UMTS femtocells, a RAN-based solution. The first multi-vendor interoperability activities for Iuh femto solution have already been announced. CDMA femto standards are being published in 3GPP2 with a CN-based solution (utilising IMS) and significant progress has happened for LTE, WiMAX and UMTS-IMS versions, while GSM so far presents only pre-standard solutions.

6

Femtocell Management

Ravi Raj Bhat and V. Srinivasa Rao

6.1 Introduction

This chapter addresses the management and configuration needs of femtocells, building on a previous article by the same authors.[1]

While femtocells have the potential to enable a very high-quality wireless connectivity experience in the home or office, one of the challenges is the mushrooming numbers of FAPs in uncontrolled customer premises environments. Base stations and Node Bs in the macrocell network are controlled, managed and sometimes owned by operators. However, in femtocell environments, subscribers are expected to walk into a store, buy a FAP, sign up for a femtocell service, take the FAP home, connect it to an existing broadband access network, and start using the femtocell service; similar to and preferably simpler than today's Wi-Fi experience. One important difference is that, while Wi-Fi works on unlicensed spectrum, femtocells work in licensed spectrum, which is an operator's asset. So, operators need to have full control over that critical asset (wireless spectrum), similar to what they have in the macrocell environment. Operators want to control who uses the femtocell service through the FAP and how they use it, while ensuring a zero-touch experience for the subscriber. This requires zero-touch remote provisioning and management capability. Provisioning involves initial bringing up of the FAP and the femtocell service, whereas management involves ongoing control of the FAP and femtocell service. In the International Standards Organization's (ISO) Telecommunication Management Network (TMN) terminology, this involves fault, configuration, accounting, performance and security (FCAPS) management.

As femtocells work over existing broadband access networks, it would be natural to leverage the existing FCAPS capability of broadband access networks. This involves mapping broadband access networks' FCAPS capability to meet the requirements of the Fm management interface identified in the Femto Forum reference model, as illustrated in Figure 6.1 and

---

[1] This chapter is modified and extended, by permission, from an article authored by Ravi Raj Bhat and V. Srinivasa Rao and originally published in (108).

Femtocells: Opportunities and Challenges for Business and Technology   Simon R. Saunders, Stuart Carlaw, Andrea Giustina, Ravi Raj Bhat, V. Srinivasa Rao and Rasa Siegberg   © 2009 John Wiley & Sons, Ltd

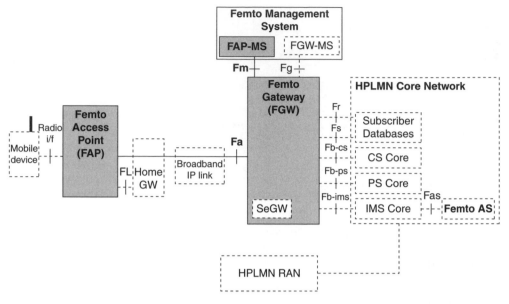

Figure 6.1   Femtocell architecture reference model. *Reproduced by permission of Femto Forum Ltd*

explained more fully in Chapter 5. The Femto Forum has made great strides in working with the Broadband Forum (previously known as the DSL Forum) in standardising the Fm interface by leveraging the existing remote device configuration and management standards, particularly the TR-069 management protocol (77) and the template data model for devices managed using TR-069, TR-106 (121). This covers the DSL broadband access network to a large extent and has gained significant momentum in 3GPP and 3GPP2. There is further ground to cover with cable modem and data-over-cable service interface specification (DOCSIS) broadband access technology. This is particularly evident in WiMAX, where the WiMAX Forum is recommending a choice between TR-069 and DOCSIS for the WiMAX femtocell management needs (79).

The following subsections first outline the FCAPS requirements for femtocells and then focus on explaining the existing Broadband Forum remote device management framework and how it could be adapted to manage FAPs. The section also briefly touches upon the DOCSIS operational support system (OSS) network architecture for cable modem broadband access networks and how it could potentially be leveraged to manage FAPs.

6.2 Femtocell FCAPS Requirements

This section briefly touches upon the FCAPS goals for femtocells. In the following subsections, the network management system (NMS) is used to represent the network operator's management system. In common with terminology used in standards, 'shall' is used to indicate a mandatory requirement, which must be supported by all implementations.

6.2.1 Fault and Event Management

The goal of fault and event management is to provide an appropriate event notification mechanism to identify significant events (including faults) and the management interface control for the NMS to take appropriate action on FAP, femtozones controlled by the FAP, or the subscriber locked to FAP. Some of the broad fault and event management goals are:

- FAP shall provide control to enable/disable the reporting of events to NMS, and control the frequency of reporting these events.
- FAP shall capture as much information as possible associated with an event, such as number of times of its occurrence since the start up, category of event, cause code, severity, timestamp etc.
- FAP shall be able to issue events to the NMS based on the configuration of the specific event's reporting mechanism i.e., either proactive (push to NMS without request from NMS) or reactive (on request from NMS).
- FAP shall provide backward compatibility at the management interface so that it can interwork with an NMS operating at a lower version than the FAP.

6.2.2 Configuration Management

Configuration management focuses on establishing and maintaining consistency of a product's performance and its functional and physical attributes throughout its operational life. Some of the broad configuration management goals are:

- FAP shall be identifiable with a unique identity and the NMS can associate this identity with a customer and location. FAP has to register with the NMS, which shall be able to authenticate and authorise it based on the identity credentials.
- FAP shall provide configurable/reconfigurable operational parameters on the management interface to the NMS.
- FAP shall provide standardised configuration/reconfiguration parameters at the management interface to enable interoperability with any vendor/operator NMS.
- FAP management interface shall be extensible to accommodate the future needs/ requirements.
- FAP shall be able to dynamically discover the configuration server.
- FAP management interface shall have an extensible data model for the NMS to be able to provision/re-provision new services in FAP.
- FAP management interface shall be able to identify problems in the new configuration and be able to revert to a working configuration, which was previously configured by the NMS.

6.2.3 Accounting and Administration Management

Accounting involves gathering usage statistics for femtocell subscribers to bill them for the femtocell service provided. Administration often involves administering a set of authorised femtocell subscribers by establishing user identity, passwords, access control lists, and administering the operations of the FAP.

6.2.4 Performance Management

The goal of performance management is to monitor the femtocell network to determine current efficiency, throughput, percentage utilisation, error rates and response time to proactively take care of femtocell network health issues and evolve the network for future requirements. Statistical performance trends can indicate capacity or reliability issues before they start affecting services. Performance thresholds can be set in order to trigger an alarm, which would be handled by the normal fault management process.

6.2.5 Security Management

The security management goal is to control access to femtocell network assets. This involves authentication and authorisation of the FAP to use network services, and femtocell subscriber authentication, authorisation and key management. Security management also involves administering an access control list of all the femtocell subscribers allowed to use femtocell services on a specific FAP. Security management shall also account for signalling and bearer traffic security, and security of the control management interface. Finally, security management shall account for the secure distribution and revocation of security credentials. Chapter 7 elaborates security aspects of femtocells.

6.3 Broadband Forum Auto-Configuration Architecture and Framework

The guiding principle around the Broadband Forum's auto-configuration architecture and framework is to configure the customer premises equipment (CPE, also known as the broadband network termination, or B-NT in Broadband Forum terminology) with 'zero touch' based on a predefined service configuration template located somewhere in the service provider's network, thus avoiding a costly truck roll to the customer premises and enabling a true plug-and-play experience for the customer.

Figure 6.2 illustrates the end-to-end network architecture required to achieve auto-configuration. The auto-configuration server (ACS) stores the predefined service configuration template, added and pre-validated by the service provider through its service configuration manager. The Broadband Forum's technical report, TR-046 (126), defines the DSL auto-configuration architecture and framework. This generic framework can be divided into three distinct aspects:

- Data organisation, validation and storage, which is typically described in a data model for a specific technology (e.g., TR-106 (121)).
- Process and transport protocol used to convey the configuration information from the ACS to a CPE device; this is described in TR-069 (77).
- Configuration information used in the CPE device – leading to service activation – which is device-specific.

The following sections will briefly explain data model organisation and TR-069, the CPE WAN management protocol (CWMP), followed by the specific data model developed by the Femto Forum and Broadband Forum to manage FAPs.

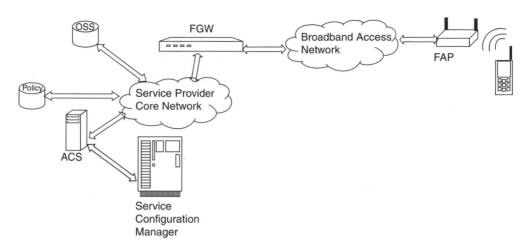

Figure 6.2   Auto-configuration network architecture

6.4 Auto-Configuration Data Organisation

A very important aspect of auto-configuration and flow-through service provisioning is the way in which configuration and service information are represented in the service provider domain. Such information needs to enable service providers to easily extend the configuration representation and quickly define new services with minimal incremental costs to roll out the new services. The configuration information representation is referred to as the *data model* and an object-oriented model is preferred for this purpose as it allows easy extensibility and fits very well with the underlying remote procedure call (RPC) methods used by protocol-independent transport mechanisms. TR-106 describes the data model template for TR-069-enabled devices. The following sections explain this template briefly.

6.4.1 Data Hierarchy

The data representation for a TR-069-capable device will always have a single root object, which will be called either *Device* or *InternetGatewayDevice*. Typically the root object contains two types of sub-elements: the common objects, applicable only to a *Device* root object; and a single *services* object that contains all the service objects associated with the specific services or applications. For *InternetGatewayDevice*, the root object will also contain the application-specific objects associated with it. A single device might include more than one service object (e.g., a device that serves both as a FAP and an IPTV set-top box might include both FAP-specific and IPTV-specific service objects) and/or more than one instance of the same type of service object (e.g., where a TR-069-capable device proxies the management functions for one or more other devices that are not TR-069 capable). Figure 6.3 illustrates the data hierarchy and how the template is defined. Figure 6.4 illustrates the TR-069 data model structure, where each box represents an object container.

Re-definition of the service object or root object over time is allowed via object version-ing with the first version starting at '1.0'. Object version is defined as a pair of integers ('major'.'minor') separated by '.', where the first integer is the major version number and the

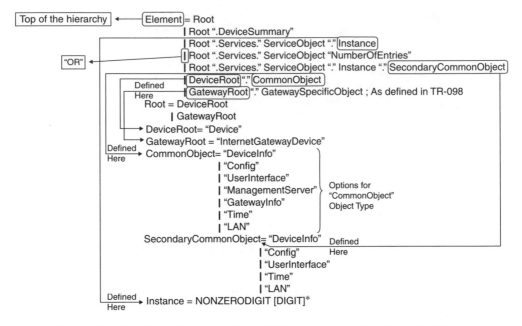

Figure 6.3   Data hierarchy. *Adapted from (121) by permission from Broadband Forum*

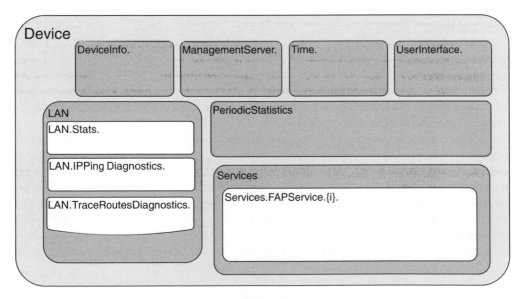

Figure 6.4   TR-069 data model structure

second integer is the minor version number. The major version changes if the subsequent version is not compatible with the previous versions, otherwise only the minor version changes.

6.4.2 Profiles

The ACS needs to be scalable enough to communicate with multiple devices with varying capabilities and potentially from different manufacturers. This variability is controlled by defining the *profiles* that express a collection of requirements associated with a given object, support for which can be explicitly indicated by the device. A device supporting a profile means that the device supports all of the requirements defined by that profile. The use of profiles allows the ACS a shorthand means of discovering support for entire collections of capabilities in a device.

A given profile is defined only in the context of a specific service object or root object with a specific major version. A profile's name must be unique among profiles defined for the same object and major version. A given profile is defined in association with a minimum minor version of a given object that includes all of the required elements defined by the profile. For each profile definition, the specific minimum version must be explicitly specified.

For a given type of service object, multiple profiles may be defined and they may have either independent or overlapping requirements. To allow the definition of a profile to change over time, the definition of every profile has an associated version number which uses a minor-only version numbering convention. All compatible changes to a profile use the same profile name but different minor versions. Any incompatible change to a profile shall use a different profile name. For every service and root object there is at least one baseline profile defined which supports the minimum requirements required for any device that supports that object.

6.5 CPE WAN Management Protocol Overview

TR-069 defines a CPE WAN management protocol (CWMP) for secure auto-configuration of CPE devices and provides other CPE management functions in a common framework, including:

- *Auto-configuration and dynamic service provisioning* of a CPE device either on initially connecting to the broadband network or later while re-provisioning or re-configuring to allow services and capabilities to change in the future. CPE devices are identified based on various criteria such as CPE vendor, model, software version, etc.
- *Software/Firmware image management* including mechanisms for version identification, file (group or single) download initiation (ACS initiated downloads and optionally CPE initiated downloads), authentication of file source, and notification of the ACS of the success or failure of file download.
- *Status and performance monitoring* of the CPE device.
- *Diagnostics* for CPE to report critical information to the ACS, which may use it to diagnose and resolve connectivity or service issues as well as to execute specific diagnostic tests.
- *Security* to prevent tampering with the management functions of a CPE or ACS, ensure confidentiality of the transactions that take place between them, allow appropriate

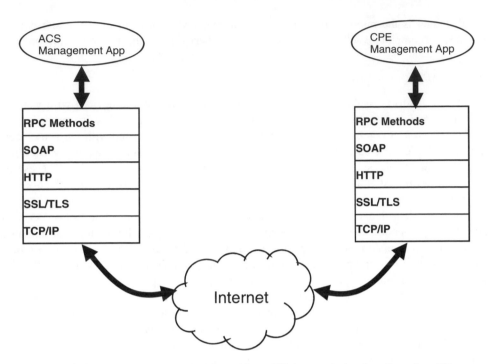

Figure 6.5   TR-069 protocol stack. *Adapted from (77) by permission from Broadband Forum*

authentication for each type of transaction, and prevent theft of service. CWMP to a large extent leverages the security services provided by underlying layers (e.g., SSL/TLS).

6.5.1 Protocol Stack and Operation

Figure 6.5 illustrates the protocol stack used in the CWMP. *CPE/ACS Management App* uses the CWMP protocol on the CPE and ACS and is locally defined by device vendors.

RPC methods define a generic mechanism by which an ACS can read or write parameters to configure a CPE device and monitor CPE status and statistics. Each parameter consists of a 'name-value' pair where 'name' identifies the particular parameter and has a hierarchical structure similar to files in a directory, with each level separated by a dot. The 'value' of a parameter is one of several defined data types. RPC methods also define a mechanism to facilitate file downloads or optionally uploads for a variety of purposes, which includes firmware upgrades or vendor-specific configuration files. File transfers can be performed by unicast (for downloads) or multicast transport protocols. Unicast protocols include HTTP/HTTPS, FTP, SFTP, and TFTP. Multicast protocols include FLUTE and DSM-CC.

RPC methods are encoded using the standard XML-based syntax called SOAP. CWMP recommends using the SOAP 1.1 protocol (127). SOAP runs over the standard HTTP 1.1 protocol and security is enabled using SSL 3.0 and TLS 1.0 specifications, which run over the standard TCP/IP protocol. The CPE device acts as the HTTP client and the ACS acts as the HTTP server.

6.5.1.1 Protocol Operation

The CPE device establishes a connection toward the ACS when it boots up for the first time or at a suitable trigger such as a change in the URL of the ACS. The ACS can be configured in the CPE or alternatively the CPE can discover the ACS using DHCP. After identifying the CPE (using the CPE vendor, model, software version, or other criteria), the ACS configures the writeable parameters in the CPE using RPCs. All of these configurations take the form of request/response and form a transaction; hence a series of transactions might be required to configure the CPE based on the number of parameters to be configured. All these transactions can happen in a single TCP/IP connection or they could span across multiple TCP/IP connections.

An event is an indication that something of interest has happened that requires the CPE to notify the ACS via an Inform request. The CPE must attempt to deliver an event at least once. If the CPE is not currently in a session with the ACS, it must attempt to deliver events immediately by initiating a session with the ACS, otherwise it must attempt to deliver them after the current session terminates.

All transaction sessions must begin with an Inform message from the CPE to the ACS contained in the initial HTTP POST. This will be used to initiate a set of transactions and to communicate the limitations of the CPE with respect to message encoding. The session terminates when both the ACS and CPE have no more requests to send and no responses pending from either the ACS or the CPE. Only one transaction session shall exist between the CPE and ACS at a time. Please refer to Figure 6.6 for an example transaction session.

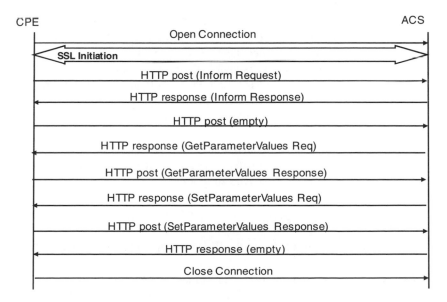

Figure 6.6   Transaction session example. *Adapted from (77) by permission from Broadband Forum*

6.6 FAP Service Data Model

The Femto Forum has supported definition of the FAP service data model for WCDMA along with the Broadband Forum's Working Group 3. Reference (134) outlines the current working draft (submitted as a contribution to the Broadband Forum) of the FAP service data model as defined by the Femto Forum. The FAP service data model is being defined based on the TR-106 template (121) and will be transported over the CWMP protocol defined in TR-069. Much of the following text is part of the current working draft. This working draft is evolving very rapidly and at the time of publication of this book FAP service data model specification in the Femto Forum and Broadband Forum (122) may be more current than the contents of this section.

   Figure 6.7 illustrates the FAP service data model, which is defined in (134) as a service object and called *FAPService*. *FAPService* is a container for a collection of three broad categories of management objects that cover all the aspects of FAP management. The base *FAPService* object includes parameters to identify whether the FAP service instance is enabled and the number of RF instances supported. The following sections briefly describe these management objects; refer to reference (134) for details on the parameters.

6.6.1 Control Object Group

This group covers all the objects and parameters required to control the FAP's operation including:

- *Capabilities* object contains parameters characterising the capabilities supported by the FAP such as whether it is equipped with GPS; its maximum transmit power; whether it supports GSM, HSDPA, HSUPA, FDD/GSM; etc. Currently it supports objects for only the UMTS technology. Others are for further study.
- *FapControl* object contains state management of the FAP and associated control of the FAP done by the network side. The base *FapControl* object contains parameters characterising the general FAP state such as whether the FAP is enabled, whether FAP administration is locked, whether the RF transmitter is enabled, and the type of system supported (e.g., WCDMA). Device state management is based on X.731. Currently it supports objects for only the UMTS technology. Others are for further study. In the future, new objects for additional RF technologies such as cdma2000 and GSM are expected to be added to the *FAPControl* base object.
- *AccessManagement* object ensures management of subscription-based information. The base object contains parameters to identify whether access control list (ACL), closed subscriber group (CSG), and local IP access (LIA) are supported.

6.6.2 Configuration Object Group

This group covers the aspects to configure the FAP for proper operation, including:

- *CellConfig* object contains configuration management of FAP functions and protocols. The base object contains parameters to identify the type of protocol supported (e.g., WCDMA). In addition, this object contains:

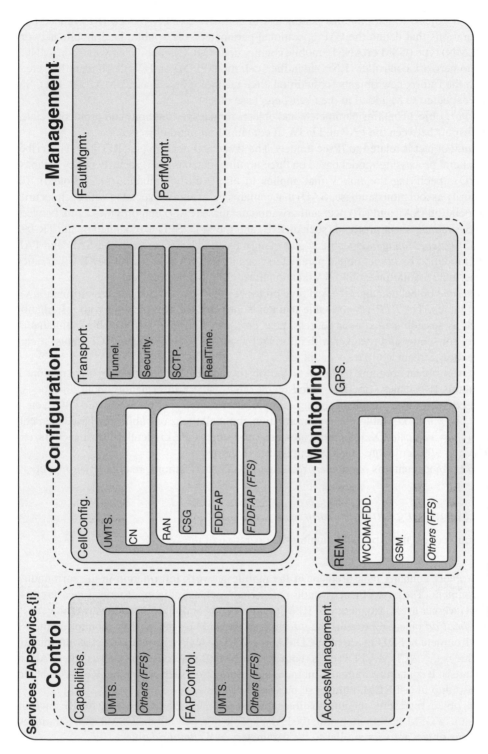

Figure 6.7 FAP service data model structure. *Reproduced by permission from Broadband Forum*

- *UMTS* object, which contains parameters to enumerate CN/RAN/cell (RF) related configurations that define the FAP operational parameters, e.g., public land mobile network (PLMN) type (GSM or ANSI), mobile country code (MCC), mobile network code (MNC), radio network controller (RNC) identifier, cell mode (FDD or TDD), cell identifier, etc.

  In the future, new objects for additional RF technology such as CDMA2000 and GSM are expected to be added to the *CellConfig* base object.
- *Transport* object contains parameters and objects to manage functions and protocols related to transport between the FAP and FGW. It contains four main objects:
  - *Tunnel* object is related to IPsec tunnels. The IPsec architecture (see RFC4301) describes a general processing model based on three nominal databases: (i) security policy database (SPD) specifying the policy that applies to all IP traffic (inbound or outbound); (ii) security association database (SAD) that contains parameters associated with each security association (SA); and (iii) peer authorisation database (PAD) that provides a link between an SA management protocol (such as IKE) and the SPD. SPD is modelled by the TR-098 (125) *QueueManagement* object, and mainly by the classification table. SAD and PAD are modelled by *tunnel* object parameters such as IKE SA peer address, IKE SA creation time, IKE SA lifetime, child SA creation time, child SA lifetime, etc.
  - *RealTime* object manages the real time protocol (RTP) session and stream information via two tables. The *RTP session* table maintains an entry for each session with information such as session status, peer address, peer port, etc. The *RTP stream* table maintains an entry for sender and receiver with information such as stream status, RTCP status, stream direction, stream lost packet count, etc.
  - *Security* object contains parameters and objects to manage security key information. It contains two tables: (i) The *shared-secret* table gathers information about all types of shared-secret-based credentials (e.g., simple shared secrets, UICC, emulated UICC, etc.); (ii) The *public key* table gathers information about all types of public key based elements (e.g., raw key pairs, X.509 certificates, etc.) and stores CPE credentials, trust anchors, etc. Basic X.509 certificate management is also supported.
  - *SCTP* object contains parameters relating to SCTP associations, remote IP address etc.

6.6.3 Monitoring Object Group

This group covers the aspects to monitor the operation of the FAP:

- *REM* object contains measurement of the mobile network information in the surrounding environment. The base object contains parameters such as radio environment measurement (REM) trigger event, frequency of REM trigger, etc. In addition, it contains two objects:
  - *WcdmaFdd* object represents the information gathered by the FAP by monitoring the RF environment in FDD mode of WCDMA in a WCDMA system. The collected information includes the WCDMA FDD cells (both regular NodeB macrocells as well as other FAPs in the area). It contains parameter-enumerating information such as whether REM is enabled, timestamp of last REM, number of measured cells, etc.
  - *Gsm* object represents the information gathered by the FAP by monitoring the RF environment in a GSM system including the FAP capable of receiving the GSM band. Parameters for this object are very similar to the parameters in *WcdmaFdd* object.

- *Gps* object contains GPS-derived location information (such as latitude, longitude, last measured time, etc.) when the FAP contains a GPS receiver.

6.6.4 Management Object Group

This group covers the aspects to manage the operation of the FAP:

- *Fault Management* object maintains the events related to faults, its history, events which are pending for delivery, and additional information such as event type, probable cause of the event, perceived severity of the event, authentication credential to upload specific event file, etc.
- *PerfMgmt* object contains parameters relating to service monitoring file management for uploading of performance files to a designated file server.

6.7 DOCSIS OSS Architecture and Framework

As noted earlier, DOCSIS is one of the approaches to be supported in WiMAX femtocell standards. Figure 6.8 illustrates the DOCSIS network architecture. The cable modem (CM) connects to the operator's hybrid fibre/coax (HFC) network and to a home network, bridging packets between them. Many CPE devices (e.g., home routers, set-top devices and personal computers) can connect to the CM's LAN interfaces. The cable modem termination system (CMTS) connects the operator's back office and core network with the HFC network. DOCSIS network leverages existing IP and related protocols for provisioning and network management.

With regard to provisioning, the dynamic host control protocol (DHCP) servers provide the CM with initial configuration information, including the device IP address(es), when the CM boots. The configuration file server is used to download configuration files to CMs when they boot. This download is achieved either using trivial file transfer protocol (TFTP) protocol or configuration-file-based SNMP encoded object and SNMP set operation. The secure software download server is used to download (via TFTP) software upgrades to the CM. Prior to secure

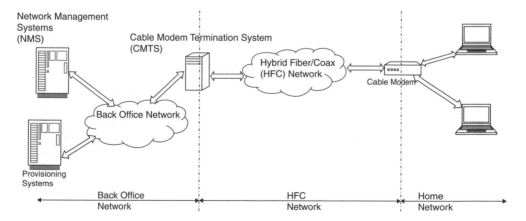

Figure 6.8   DOCSIS network architecture

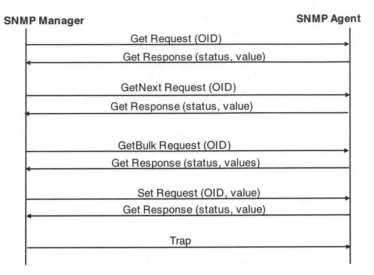

Figure 6.9   SNMP protocol data unit (PDU) flows

software download, CM should have X.509 compliant code-verification-certificate (CVC) information.

With regard to network management, the simple network management protocol (SNMP) manager allows the operator to configure and monitor SNMP agents, at the CM and the CMTS. The syslog server collects messages pertaining to the operation of devices. The Internet protocol detail record (IPDR) collector server allows the operator to collect bulk statistics in an efficient manner.

As DOCSIS primarily uses SNMP for provisioning and management, it is worthwhile to delve into SNMP basics. Two key elements in an SNMP network are:

- SNMP agent – software modules that collect and provide the management information in a network element when requested.
- SNMP manager – software modules that query SNMP agents to get management information.

The SNMP agent typically resides in managed devices such as CMs and the SNMP manager typically resides in network management stations. Management information is formally termed as managed objects and a collection of related managed objects (MO) is termed as a management information base (MIB), which is defined using a hierarchical namespace in a tree structure. Each MO has a unique object identifier (OID), identifying a variable that can be read or set via SNMP. MIBs are represented using abstract syntax notation (ASN.1). The MIB hierarchy can be depicted as a tree with a nameless root, the levels of which are assigned by different organisations. The top-level MIB OIDs belong to different standards organisations, while lower-level object IDs are allocated by associated organisations. This model permits management across all layers of the OSI reference model, extending into applications such as databases, e-mail and the Java EE reference model, as MIBs can be defined for all such area-specific information and operations. Managed objects are typically classified into two data types:

- Simple data types:
  - *integer*: signed integer with range of (-2^{31}) to $(2^{31} - 1)$;
  - *octet strings*: ordered sequence of 0 to 65,535 octets.
- Application-wide data types:
  - *network address*: represents address from a particular protocol family (IPv4/v6);
  - *counters*: non-negative integers which can increase until they reach a maximum value and then roll over to zero. It can be 32- or 64-bit based on the SNMP version;
  - *gauges*: non-negative integers that can either increase or decrease between specified minimum and maximum values;
  - *time ticks*: time since some event, measured in hundredths of a second;
  - *opaques*: arbitrary encoding that is used to pass arbitrary information strings (e.g., configuration file) that do not conform to the strict data typing used by the structure of management information (SMI);
  - *integers*: signed integer valued information;
  - *unsigned integers*: unsigned integer-valued information, which is useful when values are always non-negative;
  - *bit strings*: comprise zero or more named bits that specify a value.

The SNMP protocol operates at the application layer (layer 7) of the OSI model. Typically, SNMP uses UDP ports 161 for the agent and 162 for the manager. The manager may send requests from any available ports (source port) to port 161 in the agent (destination port). The agent response will be given back to the source port. The manager will receive traps on port 162. The agent may generate traps from any available port. SNMP uses the following protocol data units (PDUs):

- *Get Request*: used to retrieve a piece of management information.
- *GetNext Request*: used iteratively to retrieve sequences of management information.
- *Get Response*: used by the agent to respond with data to get and set requests from the manager.
- *Set Request*: used to initialise and make a change to a value of the network element.
- *Trap*: used to report an alert or other asynchronous event about a managed subsystem. In SNMPv1, asynchronous event reports are called traps while they are called notifications in later versions of SNMP.
- *GetBulk Request*: a faster alternative to using multiple GetNext requests, to retrieve sequences of management information. This was added in SNMPv2.

Overlaying femtocells onto the DOCSIS cable modem broadband access network will tie the FAP to CM. The FAP service data model explained in the previous section will have to be defined as a FAP-specific MIB that can be managed using SNMP PDUs. SNMPv3 provides authentication and encryption capability to enable secure provisioning and management of FAP over DOCSIS cable modem infrastructure.

6.8 Conclusions

The CPE WAN management protocol (CWMP) defined in TR-069 and the data model template defined in TR-106 give a very good base on which to build FAP auto-configuration, service provisioning and management functionality. The Femto Forum, working with the Broadband Forum and 3GPP, made very good progress in defining the data model for FAPs and completed the definitions for release 8 of the 3GPP standard in the first quarter of 2009, paving the way for interoperable FAPs and widespread femtocell service rollouts soon thereafter. There is further ground to cover on working with DOCSIS to define an MIB for FAP. This MIB can then be used along with the SNMP protocol to manage FAP over cable modem broadband access networks.

As a result of these efforts, we can expect 2009 and 2010 to herald broad carrier adoption and mass femtocell deployment across developed telecommunications markets worldwide. By unshackling consumers from landline phones in their homes, femtocells will help drive a new wave of subscribers, usage and applications consisting of media-rich content designed for smart, hand-held devices. The dawn of the next era of personal computing and communications is upon us.

7

Femtocell Security

Rasa Siegberg

7.1 Why is Security Important?

Assessing the importance and motivation for providing security in a new telecommunications paradigm, such as the femtocell system, can be approached from two rather distinct vantage points: those of continuity and those of contained change.

7.1.1 Viewpoint: Continuity

The continuity vantage point is really the point of view of the service-enjoying end-user – the Jill or Joe to whom the femtocell system provides indoor coverage, better cellular data rates and the possibility to enjoy the value-added services the carrier chooses to provide. While the innards of the mobile network change and evolve, new access technologies are phased in and out, and the phones themselves change, there is – and must be – a single continuation of trust shared between Jill or Joe and their mobile network operator. This trust is in part created by the uncompromising level of security the mobile network operator provides – regardless of the employed technologies or network architectures. The end-customers of a successful mobile network operator will always take for granted that their communications are safe from illegitimate wiretapping or other forms of compromise. Security in this respect is then:

- **The privacy of communication**: neither calls nor data can be illegally eavesdropped upon.
- **The integrity of data transferred**: voice and data services provide adequate levels of service to carry messages to their destination without modification – intentional or unintentional.

In short, changes to the mobile network architecture, such as the femtocell system represents, cannot be allowed to decrease the level of confidentiality or integrity protection to the end-customers.

Femtocells: Opportunities and Challenges for Business and Technology   Simon R. Saunders, Stuart Carlaw, Andrea Giustina, Ravi Raj Bhat, V. Srinivasa Rao and Rasa Siegberg   © 2009 John Wiley & Sons, Ltd

7.1.2 Viewpoint: (Contained) Change

The other angle from which to inspect the security aspect of the femtocell system is to take a look at the overall mobile communications network with the eyes of a system architect. Such an inspection quickly reveals that the introduction of the femtocell system changes one of the fundamental underlying assumptions of mobile communications, and while doing this it also introduces new threats into the security landscape of the overall system. This fundamental change is that femtocells alter the borders between the domains of control of the end-users and the network operator; while the mobile network architectures thus far have placed only the terminals at the hands of the customers, the femtocell paradigm extends parts of the radio access network into the private domains – homes and offices – of the end-users.

This is a remarkable change and it carries equally noteworthy implications to the security threat model of the femtocell system. From a 'contained change' perspective, the changes described in the network architecture must not be allowed to decrease the overall security of the mobile communications network.

Regardless of the approach one takes in looking at the security of the femtocell system, addressing the security implications of the femtocell system is equally important.

7.2 The Threat Model

Assessing the security needs of any system is generally best approached by assessing the perceived threats to the system: the threat model, as illustrated in Figure 7.1. This section seeks to group the threat model of the femtocell system into categories and then observe the threats that those categories contain.

The simplest categorisation of threats within the femtocell threat model is to observe the model from the point of view of the would-be attacker. This way the threat categories can be divided into:

- threats from attackers that are 'outside' the femtocell system; and
- threats from attacks by 'insiders' that work within the system.

An example of an 'outside attacker' would be a third party that mounts attacks on the communications links between the end-users and the carrier's network. Such an attack could

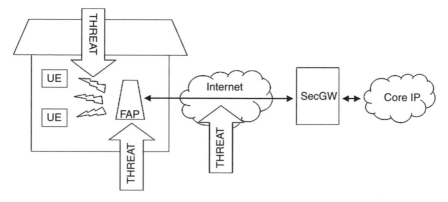

Figure 7.1   A simplified threat model for the femtocell system

be an attempt to gain illegitimate network access via the femtocell (by faked authentication or 'piggybacking' on legitimate communication) or an attempt to eavesdrop on transmitted data or voice (by capturing and analysing data that flows between peers).

An 'insider' attack scenario, on the other hand, could be to subvert the femtocell system or parts of it to work in a fashion not intended or prohibited by the network operator (for example to overcome the geographical location fix of the access point, or to use the cellular data capabilities of the access point for illegitimate purposes).

The following sections present an overview of some of the possible attacks that are presented by the threat model. This list is not comprehensive (and not intended to be such) but serves as a motivation to the subsequent sections that discuss some of the identified countermeasures and security technologies. For a more thorough analysis of the various threats to the femtocell system, the reader is directed to the relevant documents from the 3GPP, for example 3GPP TR 25.820 (46) and TR 33.820 (84) and also to (96).

7.2.1 Threats from 'Outsiders' – Third Parties

As stated earlier, an attacking 'outsider' or a third party is someone who can only observe the system from the outside (he/she has visibility into the system's externally visible interfaces). In practice for a femtocell system this would mean that the attacker is able to observe, capture and analyse the data streams that flow in and out of the femtocell access point (FAP) either over the radio frequencies used or via the data link that carries the IP connections to the carrier network. The attacker does not have physical access to the access point or to the carrier-hosted components of the system, but can be deemed to have available a user equipment (UE) that is able to attach itself to the FAP (depending on whether or not the FAP successfully enforces the 'closed user group' or not).

The lack of physical access to the system leaves the outsider with at least the following options or 'avenues of attack':

- attempting to eavesdrop on the cellular radio;
- attempting eavesdropping on the IP backhaul;
- attempting to gain illegitimate cellular network access via the FAP;
- attempting to gain illegitimate IP access to the carrier's core network;
- mounting a denial of service (DoS) attack on the access point or the carrier's core network components.

While this half of the threat model is mostly concerned with the threat of attacks that would be perpetrated towards the access points and the links between the base stations and their communicating peers (mobile terminals on the radio link side and the SecGW on the backhaul side), the next section will move on to observe security from the viewpoint of the access point device itself.

7.2.2 Threats from 'Insiders' – Device Owners

Deployment of the femtocell access points at the customer/end-user premises will subject the access points to an entirely new category of threats compared to the base station generations to date. So far a working assumption on all radio access network equipment has been that the devices themselves have physically enjoyed a degree of security due to physical access control at base station sites. With the femtocell access point deployments, this control has been relinquished, and the end-users now have full and unrestricted physical access to the base station (and radio network controller) equipment.

The 'inside' attacker in this threat model is then a person who has physical access to a femtocell access point, and most likely also a phone/terminal that is able to attach to the network via that access point. The attacker also has (or has access to a) legitimate subscription to the carrier's femtocell system. The motivation of the attacker can range from curiosity to malice, but the 'avenues of attack' in both cases are to:

- tamper with the device to make:
  1. illegitimate software modifications
  2. illegitimate hardware modifications
- pry on the data stored on the device (carrier's configuration data, usage statistics, possible billing data, etc.).

The history of tamper protection for CPE (customer premises equipment) devices is somewhat depressing from the security point of view. Tamper resistance schemes have so far been deployed in a wide variety of devices (ranging from mobile phones to home entertainment systems) – yet the degree of success of these systems has sometimes been called into question as many of the schemes have been eventually compromised: SIM-locked mobile phones are regularly 'unlocked', media protection in gaming consoles has not stopped software piracy, region coding of media players seems easy to overcome. The ongoing struggle between the anti-tamper and integrity protection methods and the persons seeking to overcome them seems to highlight this as one of the most pressing issues to address in the femtocell access point security design and vendors are rising to the challenge in a variety of ways.

7.3 Countering the Threats

The previous section observed our simplified threat model of the femtocell system and outlined some of the possible attacks that the system may face once deployed. In this section we will take a look at the various methods, tools and protocols that are currently available for countering the identified threats.

The discussion of security technologies and countermeasures is divided into the following sections based on the aspect of the femtocell system in which they can and should be used. These three aspects are:

- Authentication and encryption on the cellular radio link.
- Authentication and encryption on the IP backhaul.
- Platform security at the access point device.

7.3.1 Radio Link Protection

The most obvious communication link to be attacked in the femtocell system is the air interface between the subscribers' mobile phones and the access point. Due to the physical characteristics of the radiowaves, the communications are visible to any device capable of listening on the appropriate radio frequency. This is of course an aspect of the mobile networks that has over

Table 7.1 Overview of cryptographic algorithms in the air interfaces

Mobile communication system	Encryption
GSM	The *A5/1* & *2* stream cipher
3G (WCDMA, CDMA2000)	The *Kasumi* (aka A5/3) and
	AES (Rinjdael) block ciphers
LTE	*AES* and *SNOW* block ciphers
WiMAX	*AES* or *3DES*

the years received much attention, and the cellular radio protocols of today and tomorrow incorporate sophisticated security mechanisms that facilitate the strong authentication of the mobile terminals and subscribers as well as voice and data encryption for privacy and integrity protection.

From the point of view of the security of the femtocell system, the leveraging of these existing and well-specified security solutions echoes one of the key attributes of femtocells as listed in Section 1.4; by adopting the *mature and proven mobile technologies* the femtocell system's air interface enjoys an equal level of security to that of the macrocell mobile system.

7.3.1.1 Encryption on the (FAP to UE) Air Interface

The aforementioned mobile communications protocols (GSM, UMTS, LTE, CDMA, WiMAX and others) all utilise symmetric encryption algorithms for ensuring the privacy of the voice and data traffic between the UEs and the base station. These security measures share the common characteristic of most encryption algorithms in being somewhat computationally expensive, a fact that carries with it some noteworthy implications for the design of the femtocell access point devices and the semiconductor designs within them.

Table 7.1 presents some of the encryption algorithms employed in some well-known cellular radio protocols.

7.3.1.2 Challenges and Solutions

In a macrocell network, or 'traditional' mobile network, there are bound to be large numbers of concurrent calls per base station. The network equipment that serves the mobile subscribers is generally located in operator-controlled premises, and has relatively lax requirements on computational efficiency, power consumption and physical dimensions. The inclusion of adequate computational resources to be able to handle the overhead imposed by the employed security protocols is generally not an issue.

While the facilitation of cryptography for the encryption on the air interface with adequate computational resources is not problematic for the macrocell base stations, the same does not hold true for a femtocell base station that operates on relatively low-end, low-cost computing hardware, and under strict restraints for both physical dimensions as well as for resources such as available memory, power etc.

One of the design challenges faced by the developers of the access point hardware platforms is how to efficiently facilitate the required level of computing power for encryption and decryption of the traffic in a femtocell access point. The present hardware platform designs available have approached this problem by including in the access network specific cryptographic capabilities in the semiconductor designs used in the access point platforms, for example, most 3G femtocell designs offer hardware support for the *Kasumi* algorithm integrated into the semiconductor designs that power the device. This design choice allows the use of a less computationally powerful (and hence less power hungry) and more cost-effective general-purpose CPU to be used at the access points.

7.3.2 Protecting the (IP) Backhaul

Possibly the most revolutionary aspect of the femtocell system is the way it leverages the residential broadband and/or public IP networks to connect the FAPs to the carrier's core network. By extending the carrier network to utilise and encompass elements and technologies not under the direct control and supervision of the mobile operator, this departure from a 'traditional' mobile network architecture also introduces some security issues for the femtocell systems. The following section considers the use of the public IP networks for the 'femtocell backhaul', the security implications of it and the security technologies for addressing the issues therein.

7.3.2.1 Leveraging IPsec and IKEv2

The use of the residential broadband and public IP networks (i.e. the Internet) as a component of the femtocell system naturally requires that the connection be made as secure and trustworthy as if it was a part of the carrier's network proper. This requires that:

- The FAP and the network must be able to *mutually authenticate* each other for the FAP to become part of the carrier's network.
- Upon successful authentication the FAP and the carrier network must *create a secure connection* between them.

To accomplish the above, the femtocell system leverages standardised and mature protocols for IP security, authentication and key exchange – the IPsec and IKEv2 protocols.

7.3.2.2 Authentication and Key Exchange with IKEv2

Standardised by the IETF (Internet Engineering Task Force), the Internet Key Exchange version 2 (IKEv2) provides a robust means to address the authentication requirements of FAPs and SecGWs. As its name implies, IKEv2 is the second version of the protocol that was designed for IPsec VPN use in the 1990s, and then standardised in a series of IETF RFC documents (the RFC 24xx series). The widespread practical usage of this protocol revealed some deficiencies that were addressed by redesign and reworking of the protocol. The outcome of this effort was written as a new series of standards for the IKEv2 protocol (in the RFC 43xx series) (64). The redesign included many notable enhancements and the latter standard is:

- more efficient with less overhead: four messages passed between peers (compared to six or nine in IKEv1);
- more secure: the passed messages are authenticated to prevent a DoS attack with fake messages;

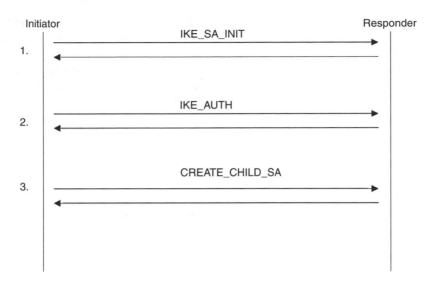

Figure 7.2   IKEv2 message exchanges to establish an IPsec tunnel

- more flexible in negotiation (address and port ranges, lists);
- simpler, which results in better implementations.

This protocol has been adopted by 3GPP as the primary method of setting up secured IPsec connections in a number of roles ((115), (116), (117)).

IKEv2 is a flexible protocol that supports a wide and varied set of use cases with support for many actual authentication methods. The IKEv2 protocol supports the mutual strong authentication of *communicating peers* over PKI certificates, shared keys or with the use of the various methods of the *extended authentication protocol (EAP)*. The use of EAP within IKEv2 provides the means for an even wider set of authentication protocols, such as the EAP-AKA and/or EAP-SIM that leverage the existing authentication back-ends and AAA servers in the telecommunications use cases. The support for EAP is a key feature of the integration of IKEv2 with many existing authentication schemes or systems, for example the UICC-based authentication mechanism utilised by the mobile telecommunications services.

Figure 7.2 and the following outline explain the IKEv2 message exchange needed to establish an IPsec tunnel between two peers. The message flow is:

1. The first pair of IKEv2 messages (or packets) carries the IKE_SA_INIT exchange. The purpose of this exchange is to negotiate a mutually agreeable set of cryptographic parameters, nonces (numbers used once), and to complete a Diffie–Hellman exchange between the peers.
2. The second pair of messages carries the IKE_AUTH exchange. In this exchange the peers authenticate the previous messages, present to each other their identities (and in some cases certificates), and establish the first CHILD_SA. These messages are partially encrypted and integrity protected to hide identities of the peers from possible eavesdroppers. This phase of the negotiation can also be used to carry payloads such as the IKE configuration

payload, that allows for the SecGW to allocate and send the FAP networking configuration data (such as an allocated IP address to be used as a 'source address' when sending packets to the carrier core network).

3. The last exchange consists of a single request/response pair, and it can be initiated in either direction after successful completion of the previous exchanges. This exchange is only needed in case the initial negotiation requires the creation of more than a single IPsec SA. The CREATE_CHILD_SA is also used for the eventual rekeying of an active but soon to expire IPsec SA.

So upon successful negotiation, identification, and authentication of both parties the IKEv2 protocol generates symmetric keys for the actual IPsec 'tunnel' and sets up the 'tunnel' for further secure communication.

There are some considerations for IKEv2 authentication that are of particular interest in the context of using IKEv2 in the femtocell system. In particular, there is a question as to whether to authenticate a user, a subscription, a device or all three.

The IKEv2 is a 'peer-to-peer' protocol – it works between two communicating parties of equal importance (as opposed to 'client-server' protocols such as TLS, for example). The identities exchanged are typically either tied to a particular person or an entity or a particular device. In the 3GPP context, IKEv2 is most often specified to be used in conjunction with the EAP methods (EAP-SIM, EAP-AKA) that leverage the UICCs employed in the telecommunications network as the authentication credentials. Using the UICCs has many benefits:

- As a hardware-enabled and robust authentication method the EAP-SIM and EAP-AKA provide a high degree of security.
- The authentication back-ends for these methods exist, and are deployed and field-proven.
- The UICCs themselves provide a method for attaching the subscription to a customer that is verified to exist and be credit-worthy.
- The logistics chains and integration to operators' business (sales, customer care, support, etc.) are for the most part already in place for UICC-based authentication methods.

The use of SIM-based methods, does however, pose some issues as well. As the acronym SIM (subscriber identity module) highlights, the UICC were originally designed for authentication of valid subscriptions to the mobile services, and not the devices these services are used on. For a femtocell system this difference becomes evident when one considers a market in which FAP devices are readily available, but in which a carrier of a femtocell service wants to limit the service to be used on a defined set of 'carrier approved' devices. Such a scenario is naturally speculative, but it is used here to point out that it may be necessary to be able to authenticate *both* the femtocell access point device *and* the subscriber.

To facilitate an authentication process that allows for maximum flexibility the 3GPP documentation on the subject (46) suggests that technical solutions that comply with the *IETF RFC 4739 'Multiple Authentication Exchanges in the Internet Key Exchange (IKEv2) Protocol'* (118) can be used to overcome the limitations of using either the device-specific or subscriber-specific authentications alone. The combining of multiple authentications is not covered in the original IKEv2 standards, but the extensions mentioned allow for this.

The use of PKI certificates provides an alternative method for strong authentication of access points and security gateways. The use of asymmetric cryptography and certificates is a proven technology that provides strong, robust and very scalable peer authentication. This method has been used with the different IKE versions since the very beginning of the IPsec and IKE protocols and offers an attractive alternative (or complement) to the UICC-based authentication methods. An important aspect of the certificate authentication is, however, that the use of asymmetric cryptography requires that the 'certificate holding' endpoints (access points and gateways) take proper care of the private key of the entity's key pair. In the access point this assumes a way to store the private key in a secure manner that allows the key usage but negates attempts to read and copy the key. The secure storage requirements this imposes are further discussed in Section 7.3.3 as well as in (46).

7.3.2.3 IPsec for a Secure Backhaul

IPsec, another IETF standardised protocol for securing Internet communications, is a required security element for protecting femtocell network backhaul – any communications between the access points to the security gateway. The IPsec protocol is widely used within the telecommunications systems as the privacy and integrity protection mechanism for IP packet traffic and as such was a natural selection to be used in the same role in the femtocell system as well.

IPsec protects the IP traffic as it travels over the broadband connection back to the carrier's core network. IPsec is a flexible and efficient method of providing data integrity, authentication and confidentiality. While IPsec is a complex suite of many protocols, backhaul security within femtocell networks focuses specifically on one variant, the *encapsulating security payload (ESP) in tunnel mode*, illustrated in Figure 7.3.

As a network level protocol, IPsec performs encryption and decryption of each packet that flows back and forth between two networking devices, the FAP and the SecGW.

IPsec is generally thought of as the creation of secure communications tunnels. In the case of femtocell network security, an ESP tunnel is created as a result of a successful IKEv2 negotiation, in which the communicating peers mutually authenticate each other and agree on the security parameters and key materials to be used.

As is the case with the encryption on the radio links, the IPsec protocol also involves the use of strong cryptography and implies a computational cost that must be taken into account

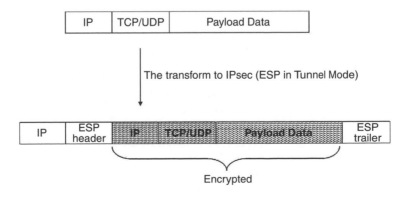

Figure 7.3 IPsec in ESP tunnel mode

Table 7.2 IPsec deployment details

Summary
Deployment model:
• IPsec (ESP protocol in tunnel mode)
• Access points always as initiators
• IKEv2 configuration payload to be used for distributing IP addresses to FAPs
• IKEv2 with multiple authentications (RFC 4739)

Femtocell access point (FAP)	Security gateway (SecGW)
1 active tunnel (2–4 IPsec SAs)	100,000s active tunnels
Low bandwidth requirements to comply with residential broadband	Extreme (1+ Gbps) bandwidth requirements to cater for the aggregated traffic of massive deployments
Media-oriented traffic patterns (small packets dominant)	Media-oriented traffic patterns (small packets dominant)

in the design of the access points and SecGWs alike. While the resource scarcity is clearly evident and becomes an issue to be solved by the hardware architectures of the access points, the same problems need to be addressed also at the SecGW. However, the motivation for the use of specialised cryptographic hardware is slightly different at the access points and the security gateways:

• At the femtocell access points the motivation for the use of cryptographic hardware is to conserve the general-purpose CPU for other tasks.
• At the SecGWs the motive and justification for cryptographic hardware is acceleration. The hardware is used to accelerate the maximum achievable encrypted throughput to the 'Gigabit-class' and beyond.

Table 7.2 presents a summary of the IPsec deployments in femtocell systems.

In many 'traditional' or Internet deployment scenarios the computational cost of cryptography of IPsec is borne by the platform CPUs. Due to a number of reasons (CPU cost and heat dissipation to mention some) this paradigm is not compatible with the femtocell model. The following section explores some of the special characteristics of the femtocell system from the perspective of IPsec use.

7.3.2.4 Special Characteristics of IPsec in the Femtocell System

While the IPsec protocol is a very mature and field-proven technology, there are some characteristics of the femtocell system that have a significant impact on the deployment of IPsec in this context:

• The commercial success of femtocell systems requires the FAPs to be affordable, which means that high-performance and therefore expensive CPUs are not a viable option. To enable the FAPs to be computationally capable of dealing with their security needs (IPsec and other), these requirements clearly prescribe the use of specialised cryptographic hardware to allow for use of low-end and low-cost general-purpose CPUs. For packet encryption and decryption to be efficiently managed FAPs and SecGWs alike will need to rely on cryptographic hardware for offloading the cryptographic calculations. The offloading models to be employed will differ but the presence of cryptographic hardware is attractive to take the burden of crypto away from the CPU, to allow it to retain the processing cycles necessary for the other associated tasks (managing the radio links at the FAPs, O&M etc. in both).

- Due to the real-time and media orientation of the UEs, the IP traffic patterns that the femtocell access points and gateways will need to process have higher proportions of small IP packets, which impose a higher relative overhead to the IPsec traffic (as the relative amount of 'overhead bytes' is higher in small (less than 100 byte) packets than large (1000+ byte) packets). This increase in overhead further highlights the need to efficient and resource-conservative security protocol implementations.

7.3.3 Device Integrity – Tamper Resistance

While the first two categories of threats in the femtocell threat model were primarily concerned with the protection of the end-users' data against eavesdropping on the communication links of the femtocell system, the last identified threat category shifts the focus to the protection of the carrier's or operator's data stored on the access point device itself.

From a technical and IP routing point of view the femtocell access points are gateways that guard and provide access to the carrier core IP network and the carrier's radio network. Because of these connections *and* being a customer premises device, the femtocell access points will be subject to attacks that would not be easily possible to mount towards the macrocell equipment. The motivation for these attacks may range from rather innocent engineering curiosity to definite malice, but regardless of the motives, the femtocell access points will require a degree of protection and have capability to resist tampering, so as not to become the weak link in the femtocell system's security.

The aim of the platform security features in femtocell access points is to make such attacks infeasible or ineffective.

Typical attacks that the femtocell access point devices might be subjected to are:

- Manipulating a FAP to boot with illegitimate operating system and/or software (i.e. 're-flashing attacks').
- Removing and/or altering the authentication tokens within the FAP by physical intrusion.
- Using valid (removable) authentication credentials in a manipulated FAP.
- Illegitimate modification or update of software and/or configuration.
- Breaking the 'location fix' of a FAP (i.e. taking the FAP outside the operator's license coverage).
- Tampering with radio resource management features of the device.
- Prying on the sensitive carrier data on the device (configuration data, usage statistics, subscriber info).

Again for a more thorough discussion and explanation of the various threats and attacks towards the integrity of the FAP devices, the reader is directed to the document *3GPP TR25.820* (86).

In order to realise a robust level of device integrity and security and to prevent FAPs from falling victim to the fate of many past CPE devices, the platforms these devices are architected upon must be able to provide robust low-level security features and capabilities.

Research and study on the platform security features (119), (120) (for mobile platforms rather than femtocells, admittedly) seem to suggest that the most effective way of protecting the integrity of a platform is realised with combined hardware and software security solutions. Such solutions leverage a closed 'system within a system' that is isolated from the 'open' system components and is used for security-sensitive tasks such as cryptographic signature verifications, key generation, encryption or decryption.

There are some low-level platform security features that can – when implemented efficiently and robustly – be used as a basis for countermeasures against the attacks that threaten the platform integrity of femtocell access points. These 'platform security building blocks' include:

- **Secure boot functionality** – The device will be allowed to boot only software images or accept data that is endorsed or signed by the manufacturer (or a trusted third party). In general terms this can be realised by a hardware-assisted boot process in which the images are verified by hashes[1] and digital signatures prior to executing them at boot time. Implementing secure boot as a part of a device architecture is closely connected to other device management and maintenance tasks, and should be considered as a part of a bigger picture; the secure booting procedure needs to be compatible, for example, with remote (and possibly 'over the air') device firmware upgrade mechanisms, to allow device software images to be altered post-deployment.
- **Runtime integrity protection** – While the device is running its operating system and other software, there must be a way to ensure that the critical components of the executed software (operating system kernel, etc.) cannot be altered. An attacker that is able to insert code (a virus, worm or other kind of malicious software) into the device must not be able to change the original software without the change being detected.
- **Secure storage** – The access point must be able to provide storage within a 'cryptographic safe' that is only accessible to a holder of a valid key (the operator or a party authorised by the operator). The requirements for access points call for the configuration data for the radio (radio configuration data, encryption keys, identity material, and operational statistics) to be stored within the access point itself. Other data that may require protection may be the billing-related information possibly stored on an access point. In order to achieve this data protection carrier assets must be stored in a robustly protected 'cryptographic safe' within the device.

The 'building blocks' described above can be thought to form what is sometimes referred to as a *trusted execution environment (TEE)* or a *trusted environment (TrE)* (46). In many significant ways the TrE is an important component of the access point architecture, as it provides the *'device internal'* security upon which the other, *external*, security features depend on – for example, the authentication between the FAP and the carrier network is done using credentials that are stored within the secure storage in the TrE.

To be able to provide the services required of it, the TrE requires capabilities to:

- **Maintain and access device identity**: the root of trust within an access point device needs to be a physically unalterable device identity (a 'HUK' or *hardware unique key*). The TrE is the sole component in the access point that has access to the HUK of the device, and is thus the guardian of the identity of the device.

---

[1] A *hash* is a procedure which takes an arbitrary block of data and returns a fixed-size bit string, the hash value, such that an accidental or intentional change to the data will almost certainly change the hash value.

- **Perform cryptographic operations:** the TrE must be able to perform within the crypto-graphic operations that are needed for higher level security functions. In practice the TrE must be capable of random number generation, cryptographic signature verification, hash digest calculation, encryption and decryption.

The platform protection features and security measures to be implemented in access point devices requires careful consideration on the technologies to be selected and implemented to arrive at a solution that provides the required level of protection at a cost that is feasible and affordable.

7.4 Conclusions

This chapter aims to point out the importance of security for the femtocell system. A well-architected and implemented system includes the security measures that counter the threats faced by the system, and thus provides the end-customers with the level and quality of service they have come to expect from the mobile network operators and telecommunications systems.

While even the greatly simplified threat model presented in this chapter may seem pes-simistic in nature, the early identification of potential threats goes a long way to mitigating the possible impact they may have for the eventual services. The existence of security threats is a fact of life with any system design – the upside in the femtocell system's case is that the tech-nologies and methods for containing and countering the identified threats exist, and through thoughtful and careful system design the risks posed by the threat model can be minimised. The technologies, protocols and methods chosen for the protection of the femtocell system need to be architected and integrated into the standards that specify the system for the market to arrive at solutions that are interoperable, optimised and cost-effective.

The various bodies that participate in the standardisation of the different aspects of the femtocell system have identified the need for a systematic design approach to the security of the femtocell system. The combined and considerable security expertise of organisations such as 3GPP, 3GPP2, IETF and the Femto Forum has been harnessed to select appropriate mechanisms and technologies and to define the security specifications for a standardised femtocell system. The ongoing study and standardisation work done in these organisations has already resulted in a number of advances in the area, and the continuing cross-industry effort holds great promise for the near future.

8

Femtocell Standards and Industry Groups

Simon Saunders

8.1 The Importance of Standards

All femtocells are rooted in standards. Their ability to provide services to the huge installed base of legacy mobile devices with no change implies full support for the existing standards over the air. This enables both rapid introduction to a very large number of customers and the ability to make use of well-proven standards to enable full mobility and access to a wide range of existing services.

However, there is additionally a compelling need for evolutions of existing standards to provide specific support for femtocells. There is a variety of motivations for such standards:

- Existing standards for macrocell deployments may not be efficient for cost-effective femtocell implementation.
- One standard can support many 'standard' approaches, which may not allow interoperability and therefore make system integration difficult and costly for operators.
- Each operator network may desire to implement femtocells from many vendors, in a similar fashion to the way mobiles are supplied. This makes it important that, for example, a single femto gateway can support femtocells from multiple vendors in a standards-based manner.
- Standards allow products from multiple vendors to be compared and assessed in a like-for-like manner, allowing operators to procure efficiently and for vendors to deliver products to a much wider market, reducing or avoiding operator-specific customisation costs where possible.
- Customers for femtocells can benefit from a wider choice of form factors, features and integration with a wider range of other devices.
- Standards help to drive economies of scale, reducing the costs associated with deployment and helping to support a wider range of business cases.

Femtocells: Opportunities and Challenges for Business and Technology   Simon R. Saunders, Stuart Carlaw, Andrea Giustina, Ravi Raj Bhat, V. Srinivasa Rao and Rasa Siegberg   © 2009 John Wiley & Sons, Ltd

- Standards avoid market fragmentation by recognising a class of device with wide availability and market acceptance.

However, the associated standards must be carefully determined. They need to avoid stifling the innovation and speed of development, which has been the hallmark of the femtocell sector to date. They should reflect best practice and mature requirements, fed back from practical field experience and leading operators and vendors. They need to respond fully to operator technical and business requirements, not only to current vendor capability. They should not close off options for future enhancements.

There has been a huge effort on femtocell standards in recent times, which started during 2007 and has already been successful in delivering the first standards for femtocells. The current status is summarised in the remainder of this chapter. However, this is a fast-moving area and updates to standards are appearing continuously. See www.femtocellbook.com for references to the latest standards.

8.2 GSM

Although there is significant interest in GSM femtocells, motivated by the opportunity to deliver services to a very wide installed base of GSM mobiles, there has not been any real impetus for GSM femtocell standards. This does not appear to be a particular barrier to delivering products at the present time, however. See Section 5.10 for details of the architectural approaches available for GSM. If greater demand does emerge, e.g. for applications in developing markets, there is a substantial opportunity to build standards on the work already conducted for WCDMA, such as the management framework adopted.

8.3 WCDMA

At the beginning of 2008, there was no committed plan in 3GPP to produce any standards for femtocells. There was, however, a study item – essentially a feasibility study in 3GPP parlance – looking at a variety of issues, but particularly at whether RF interference between femtocells and macrocells was manageable. This eventually produced a technical report '3G Home NodeB Study Item Technical Report' TR 25.820 (46). Note that 3GPP refers to a femtocell for WCDMA as a 'Home NodeB' and for LTE as a 'Home eNodeB'.

The study produced the following key conclusions:

- It is feasible for femtocells to operate without significantly degrading performance, provided the right interference mitigation techniques are applied, especially in the case of closed subscriber groups and interference between femtocells in high density deployments.
- Femtocells can meet regulatory requirements to radiate only when it is confirmed that this would comply with regulatory requirements relevant to the femtocell location.
- Existing mobiles can be supported without modification.
- There is scope to optimise mobility and access control procedures for existing mobiles, and an opportunity to further optimise new mobiles for enhanced performance.
- Architectural solutions for supporting femtocells within legacy core networks are feasible.

Following this outcome, it was agreed in March 2008 to create several work items to deliver femtocell standards within the 3GPP Release 8 specification. This was a momentous decision, in that it created a commitment to producing a major new standard, supporting new network nodes, architectures and interface protocols, all within the existing committed timescale for Release 8. Given that Release 8 also included the first release of standards for LTE and was timed for functional freezing in December 2008, with final completion in March 2009, this was a very ambitious target. That this target was accepted and achieved is testament to the strong desire from operators to achieve a standards-based femtocell solution in a short time scale and to the commitment of vendors to resolve differences in proprietary approaches to avoid fragmentation in the industry.

An overview of the resulting Release 8 standard is provided below, organised according to the 3GPP working group producing the standard. Indications of the likely work for Release 9 are also provided, which is approximately one year behind Release 8. All of the documents are freely available in their latest versions from the 3GPP website at www.3gpp.org and links to all of these are available at www.femtocellbook.com.

8.3.1 TSG RAN WG2 – Radio Layer 2 and Radio Layer 3 RR

RAN2 is in charge of the radio interface architecture and protocols (MAC, RLC, PDCP), the specification of the radio resource control protocol, the strategies of radio resource management and the services provided by the physical layer to the upper layers.

The main output related to femtocells is specification TS 25.367, 'Mobility Procedures for Home NodeB; Overall Description; Stage 2'. It defines how a femtocell may provide restricted access to only users belonging to a closed subscriber group (CSG), where such cells are identified by a unique identifier (CSG Identity). User equipments with CSG subscriptions have an allowed CSG list containing one or more CSG Identities which are used together with the CSG Identity broadcast by the CSG cells when conducting CSG cell selection and reselection. It also defines an 'HNB Name', which is a textual identifier that can be displayed by the phone and used to help the user manually select a CSG Identity – it is therefore similar to the 'SSID' used in wireless LAN systems. In manual selection the mobile may display a list of CSG Identities from the allowed CSG list, which is found along with the HNB Names. When the user selects an entry in the list the mobile will select the CSG cell with the best RF quality. CSG also allows the mobile to avoid scanning for cells which are not allowed, reducing scanning time and battery drain.

8.3.2 TSG RAN WG3 Architecture

RAN3 is responsible for the overall UTRAN/E-UTRAN (WCDMA/LTE) architecture and the specification of protocols for the associated interfaces.

RAN3 has standardised in Release 8 the RAN-based architecture for WCDMA femtocells. The architecture itself is specified in TS25.467, 'UTRAN Architecture for 3G HNB; Stage 2' (69). The content of this is covered in more detail in Chapter 5. The document defines the existence of the HNB (femtocell), and the HNB-GW (femto gateway) with Iuh acting as the interface between them and Iu acting as the interface with the core network. It also defines the security gateway as a separate logical entity which may or may not be physically integrated with the HNB GW and the HMS (femtocell management system), which is based on the Broadband Forum TR-069 standards. The main functions of each of the network elements are defined, including clarity as to the functional split between the elements. For example, this specifies that many of the functions which would normally be handled in an RNC are handled by the HNB, making clear that a femtocell is not simply a small base station. A number of functions are also defined, including procedures for mobile registration for both legacy and Release 8 mobiles and for registration of HNBs with the HNB-GW (which informs the HNB-GW that an HNB with a specified unique identity is available at a particular IP address and the associated geographical location and surrounding macrocells). The requirements for O&M are also included, identifying the provisioning procedure, the process of location verification, the surrounding cell information that the HNB must report, and information on the broadband connection associated with the HNB where available.

TS 25.467 also defines, at high level, the overall protocol structure for signalling over Iuh, but the details of this are provided in TS 25.468 'UTRAN Iuh Interface RUA Signalling' (70), which specifies the RANAP User Adaptation (RUA) protocol and TS 25.469 'UTRAN Iuh Interface HNBAP Signalling' (71), which specifies the HNB Application Part (HNBAP) protocol. In both cases the documents provide the detailed functions, procedures for executing these functions and the detailed protocol messages to achieve these.

8.3.3 TSG RAN WG4 Radio Performance and Protocol Aspects RF Parameters and BS Conformance

RAN4 works on the RF aspects of UTRAN/E-UTRAN, including RF system scenarios, minimum requirements for transmission and reception parameters and the test procedures for verifying conformance with these requirements.

The radio aspects of the WCDMA standard are covered in more detail in Section 4.9. RAN has made changes to two existing specifications. These are TS 25.104 'Base Station (BS) Radio Transmission and Reception (FDD)' (50) to add a new Home BS class to the existing base station classes and specify the associated requirements and TS 25.141 'Base Station (BS) Conformance Testing (FDD)' (51), which added definition of the Home BS class and associated conformance testing requirements.

An additional technical report was also created, TR 25.967 'FDD Home NodeB RF Requirements' (48), providing guidance on HNB operation for interference mitigation.

8.3.4 TSG SA WG1 – Services

SA1 works on the services and features for 3G. The group sets high-level requirements for the overall system and provides this in a stage 1 description in the form of specifications and reports.

SA1 is working towards Release 9 of 3GPP, specifying enhanced service requirements for femtocells, above and beyond those included in Release 8. These are expected to include an architectural option based on IMS plus local IP access procedures for both the Iuh and IMS approaches. These service requirements are specified in a Release 8 document TS 22.220, 'Service Requirements for Home NodeBs and Home eNodeBs (Release 9)' (83).

8.3.5 TSG SA WG3 – Security

SA3 is responsible for the security of the 3GPP system, performing analyses of potential security threats to the system, considering threats introduced by new services and systems and setting the security requirements and solutions overall.

SA3 has delivered a new technical report specific to H(e)NB security, TR 33.820 'Security of H(e)NB' (84). This includes the overall security architecture and identifies and analyses security threats. Security requirements are specified and solutions to each requirement are highlighted. See Chapter 7 for details.

Detailed security features and procedures will be defined in a new technical specification in Release 9.

8.3.6 TSG SA WG5 Telecom Management

SA5 specifies the management framework and requirements for management of the 3G system, delivering the architecture descriptions of the telecommunication management network (TMN) and coordinating all work pertinent to the 3G system telecom management.

SA5 has responsibility within 3GPP for standardising operation, administration and maintenance procedures for Home NodeB. Since the Broadband Forum's TR-069 protocol was agreed as a basis for the management protocol, SA5 is liaising with the Broadband Forum. The Broadband Forum is publishing the standard data model for management using TR-069, while SA5 are referencing this and providing the detailed requirements, procedures and flows. Their outputs are provided in a technical report TR 32.821 'SON Related OAM for HNB' which defines the OAM architecture and four new technical specifications as follows:

- TS 32.581 Concepts and Requirements (85)
- TS 32.582 Information Model (86)
- TS 32.583 Procedure Flows (87)
- TS 32.584 XML Definitions (88)

See Chapter 6 for more details.

8.3.7 Summary of WCDMA Standards

The full set of WCDMA femtocell standards documents is summarised in Table 8.1.

8.4 TD-SCDMA

3GPP Release 9 includes a RAN4 study item to analyse interference scenarios for TD-SCDMA in a similar fashion to those analysed for WCDMA. This is expected to then continue into a full set of standards for TD-SCDMA femtocells, building closely on the work conducted for WCDMA in Release 8.

8.5 LTE

Most standardisation work for LTE femtocells – or Home eNodeB in 3GPP terminology – will take place within Release 9 of 3GPP, expected for completion in March 2010. Detailed stage 1 requirements are set to be frozen by June 2009, with the detailed stage 2 and 3 architectures and procedures specified in the subsequent period. The main service requirements are specified in a Release 8 document TS 22.220, 'Service Requirements for Home NodeBs and Home eNodeBs (Release 9)' (83). LTE will build directly on the work conducted for WCDMA femtocells, with many common requirements and a high degree of commonality regarding the general solutions.

Table 8.1   Summary of WCDMA femtocell standards in 3GPP

Working group	Standard number	Standard name
RAN 2 Layer 2 and Layer 3 RR	TS 25.367	Mobility Procedures for Home NodeB; Overall Description; Stage 2
RAN 3 Architecture	TS 25.467	UTRAN Architecture for 3G Home NodeB; Stage 2
	TS 25.468	UTRAN Iuh Interface RANAP User Adaption (RUA) Signalling
	TS 25.469	UTRAN Iuh Interface Home NodeB Application Part (HNBAP) Signalling
RAN 4 Radio Performance and Protocol and BS Conformance	TR 25.820 (produced jointly with RAN 3)	3G Home NodeB Study Item Technical Report
	TS 25.104	Base Station (BS) Radio Transmission and Reception (FDD)
	TS 25.141	Base Station (BS) Conformance Testing (FDD)
	TR 25.967	FDD Home NodeB RF Requirements
SA1 Services	TS 22.220	Service Requirements for Home NodeBs and Home eNodeBs (Release 9)
SA3 Security	TR 33.820	Security of H(e)NB
SA5 Telecom Management	TR 32.821	Telecommunication Management; Study of Self-Organizing Networks (SON) Related OAM Interfaces for Home NodeB
	TS32.581	Telecommunications management; Home Node B (HNB) Operations, Administration, Maintenance and Provisioning (OAM&P); Concepts and requirements for Type 1 interface HNB to HNB Management System (HMS)
	TS32.582	Telecommunications management; Home Node B (HNB) Operations, Administration, Maintenance and Provisioning (OAM&P); Information model for Type 1 interface HNB to HNB Management System (HMS)
	TS32.583	Telecommunications management; Home Node B (HNB) Operations, Administration, Maintenance and Provisioning (OAM&P); Procedure flows for Type 1 interface HNB to HNB Management System (HMS)
	TS32.584	Telecommunications management; Home Node B (HNB) Operations, Administration, Maintenance and Provisioning (OAM&P); XML definitions for Type 1 interface HNB to HNB Management System (HMS)

In some respects it gains an advantage over WCDMA femtocells in having an opportunity for including some optimised features for femtocells already specified within the mobiles, so for example Release 8 LTE mobiles will all include closed subscriber group support. Standards for LTE are giving equal weight to both FDD and TDD modes of operation, and this is expected to extend to femtocells. This create new opportunities for operators – see Section 12.2.6.

8.6 CDMA

There is strong support for femtocells amongst CDMA operators. Indeed, the first commercial femtocell deployment was of a CDMA-based femtocell by Sprint in the United States. 3GPP2 is working on femtocell standards for cdma2000 systems. A document on system requirements, produced by the 3GPP2 TSG-S group, was published in May 2008 (89). The requirements envisage development of standards in two phases:

- Phase 1: support for residential use for legacy mobiles and femto–macro mobility, assuming the same radio interface for the femto and macro layers.
- Phase 2: enhancements (including new femto-aware mobiles) to permit femto–femto mobility, mobility between dissimilar radio interfaces and possibly enhancements to permit denser femtocell deployments.

The document includes requirements in several categories, summarised below:

- Support for synchronisation without using GPS (which is the normal means of synchronisation in CDMA base stations).
- Access control policies allowing access for particular users to be specially included or excluded on particular femtocells.

System requirements include:

- Support for all features and services included in the cdma2000 specifications.
- Determination of the geographical location of the femtocell.
- Operation over existing DSL and cable backhaul networks.

Radio requirements include:

- Mechanisms to control and minimise interference between femtos and both macros and femtos, including for co-channel operation.
- Potential to support multiple cdma2000 radio technologies and frequency assignments.

Mobility requirements include:

- Support for both idle and hard handoff in both directions between macrocells and femtocells.

Security requirements include:

- Secure communications and mutual authentication between the femtocell and the operator's network.

- Means of preventing and protecting against unauthorised access to the backhaul link.
- Over-the-air security which is at least equivalent to that provided by the macrocells.

Operation, administration, maintenance and provisioning requirements include:

- Support for automated 'plug-and-play' configuration.
- Optimisation of operating frequency PN (code) assignment and other system parameters via the OAM&P interface.
- Reporting of femtocell capabilities and alarm/error status via the OAM&P interface.
- Control of the volume of data reported to the OAM&P system to avoid overloading.

Regulatory requirements include the ability to support lawful intercept and emergency calls (in conjunction with the operator's core network). Performance requirements include the ability to scale to support millions of femtocells and to deliver minimum performance specifications modelled after the existing wide-area network specifications.

Standardisation activities based on these requirements are in progress, in particular including activities on security, to be published in a security framework specification produced by 3GPP2 TSG-S in S.S0132 and on the network architecture, to be published by 3GPP2 TSG-S in S.S0135 (78). Interoperability specifications are documented by 3GPP2 TSG-A in A.S0024 (90), including local breakout support and functionality particular to femtocells such as handin and handout. Packet data services for 1X and SO, as well as SIP/IMS-based 1x legacy voice services are specified by 3GPP2 TSG-X in X.S0059. This document also contains the basic management object information. Air interface enhancements for future femto-aware mobile devices will be documented by 3GPP2 TSG-C, either in revisions of the 1x standard C.S000x-E (91) and the EV-DO standard C.S0024-C (92)or in stand-alone documents. 3GPP2 has a concept of enhanced system selection (ESS), documented by 3GPP2 TSG-C in C.S0016-D (93). ESS describes a new provisioned data structure in the mobile device called the preferred use zone list (PUZL), which provides extensive system selection, searching and system acquisition rules to allow enhanced devices to operate in an optimised manner with femtocells. The capability is analogous to the CSG concepts in 3GPP and the WiMAX Forum, but with greater sophistication and flexibility.

Table 8.2   Summary of 3GPP2 femtocell standards

Working group	Standard number	Standard name (or description where firm title is not yet available)
TSG-S (WG1)	S.R0126-0	System Requirements for Femto Cell Systems
TSG-S (WG4)	S.S0132-0	Femto Security Framework
TSG-S (WG2)	S.S0135-0	Network Architecture Model for cdma2000 Femto-Cell Enabled Systems
TSG-A	A.S0024	Femto Interoperability Specifications
TSG-X	X.S0059	Packet Data Services for 1X and SO, SIP/IMS-based 1x Legacy Voice Services, Basic Management Object Information
TSG-C	TBD	Air Interface Enhancements for Future Femto-Aware Mobile Devices
TSG-C	C.S0016-D	Enhanced System Selection (ESS) and Preferred Use Zone List (PUZL)

8.7 Mobile WiMAX

The WiMAX Forum has an active programme of femtocell standardisation, following strong support for WiMAX femtocells by many WiMAX service providers globally. Standardisation is expected in two phases:

- Phase 1: basic femtocell with limited network features and no change in the underlying air interface and system profile. No change will be required to the underlying IEEE standards. Support for WiMAX Release 1 (IEEE 802.16e-2005) and Release 2 (IEEE 802.16 Rev2). This phase is expected to be available in the next WiMAX Release 3 (4Q09). Product availability is expected early 2010.
- Phase 2: full femtocell function with advanced network features and air interface optimisation, which will include PHY and MAC layer enhancements in the IEEE 802.16m standard. The system requirements definition document for IEEE 802.16m already includes explicit support for femtocells (94). This phase is expected to be available in WiMAX Release 4 (4Q10). Product availability is expected after early 2011.

The Service Provider Working Group of WiMAX Forum has created a set of detailed requirements for WiMAX femtocell systems (79). The document includes a wide variety of use cases for femtocells, including novel business models such as where the macro network access provider and femto network access provider are different, yet voice calls must be maintained seamlessly in the transition between networks and between individual femtocells. The document also includes requirements in several categories, some of which are summarised below.

System requirements include:

- Operation over standard broadband and independently of the presence of any macro network.
- Support for a globally unique ID for each femtocell.

Performance requirements include:

- Support for potentially millions of femtocells in a single operator's network.
- Self-provisioning with no end-user or operator input to register the femtocell with the network.
- Support for existing WiMAX air interface performance specifications except for pedestrian speed mobility.

Handover requirements include:

- Handover between adjacent femtocells and macrocells within the same network access provider.

Security requirements include:

- Air interface security mechanisms at least as for existing WiMAX base station.
- Full mutual authentication between the femtocell and core network.

- Authentication credentials in the femtocell to rely on hardware mechanisms.
- Support for locking femtocells to particular network access providers.
- Support for blocking particular femtocells.
- Support for closed subscriber groups.

Air interface requirements include:

- Avoidance of 'exclusion zones' around femtocells, and provision of an agreed service for at least 10 metres in a residential environment.
- No transmission until the location is verified by the network.

Synchronisation requirements include:

- Femtocells must be synchronised with the rest of the network.
- GPS-based synchronisation is not mandated for femtocells.
- Support for network-based synchronisation mechanisms.

Network requirements include:

- Capability to allow service providers to define specific service-level agreements for services delivered via a femtocell according to the characteristics of the femtocell and of the broadband connection.
- The femtocell should indicate multiple quality-of-service classes to be used on the broadband connection and in the broadband provider network and should support mechanisms to adapt to the performance of the backhaul.

Management requirements include:

- Support for TR-069 or DOCSIS for femtocell management.

8.8 The Femto Forum

The Femto Forum (www.femtoforum.org) is a not-for-profit membership organisation, founded in June 2007 to promote the adoption of femtocells worldwide. At the time of writing it had around 100 members, including operators (accounting for over 1.3 billion mobile subscribers) and vendors of all sizes representing all parts of the femtocell ecosystem. The Femto Forum has three main aims:

- To develop an ecosystem of companies producing the various elements of femtocell systems.
- To encourage the creation of mature and timely standards for femtocells.
- To raise the awareness of femtocells, both within the industry and ultimately to potential femtocell customers.

The Femto Forum is technology agnostic, promoting the benefits of femtocells for all mobile standards operating in licensed spectrum, including GSM, WCDMA, LTE, CDMA and WiMAX.

In order to promote these aims, the Femto Forum has agreed links with several standards bodies, including 3GPP, 3GPP2 and the Broadband Forum and also has agreements with other industry associations including the GSMA and NGMN Alliance. These agreements are typically market representation partnerships. The Forum itself does not create technical contributions to standards, but enables and encourages consensus amongst its members, who frequently create standards contributions co-signed by large groups of members. This enables standards to progress more rapidly, relieving the considerable workload of these bodies and ensuring that the contributions are mature and built on the collected expertise of the members.

8.9 The Broadband Forum

The Broadband Forum[1] (formerly the DSL Forum) is a worldwide organisation accelerating the development and deployment of broadband networks, fostering successful interoperability and managing and delivering advanced IP services to the customer. They are well known for defining standardisation and test suites for ADSL and increasingly VDSL2. They have an active programme of set-top box management work, centred on the TR-069 management protocol. As discussed in Chapter 6, this has been adopted as the prime management technique for femtocells, supplemented by data models specific to femtocell management. The data model for WCDMA has been worked on jointly by members of the Broadband Forum and the Femto Forum and incorporated in 3GPP standards, but is published by the Broadband Forum (122).

8.10 GSMA

The GSM Association[2] (GSMA) is a global trade association representing the interests of over 850 GSM mobile phone operators and over 180 manufacturers worldwide. GSMA has its own femtocell group, which has produced a number of operator-driven studies and expressions of requirements, including:[3]

- Femtocell Interference and Frequency (95)
- Security Issues in Femtocell Deployment (96)
- Management of Femtocells (97)
- Femtocell Requirements on DSL Broadband (98)
- 3G Femto Solution Implementation Guidelines (99)

8.11 Conclusions

Open, interoperable standards are a key requirement for efficient, cost-effective deployment of femtocells globally. While operators are moving forward with femtocell service launches with pre-standard, proprietary approaches, in order to gain early access to the associated benefits, they have made clear that a well-articulated roadmap towards full standards-based products is essential. All of the major mobile air interface families are now committed to delivering femtocell standards and remarkable progress has been made already. These standards are all

---

[1] www.broadband-forum.org.
[2] www.gsmworld.com.
[3] These references are available from: http://gsmworld.com/newsroom/document-library/all_documents.htm.

adopting a broadly two-phase approach: first standardising architectures to deliver services to existing user devices without modification; and then moving on to include features in user devices and the air interfaces, which will further optimise the femtocell user experience and the supported range of femtocell business models.

These standards will continue to develop. They will need to be augmented by product interoperability and certification testing to enable multi-vendor support in operator networks, giving both operators and femtocell users a better choice of functions, features and form factors at affordable prices.

9

Femtocell Regulation

Simon Saunders

9.1 Introduction

Regulation can play a key role in enabling or retarding any given industry. In broad concept for regulators, femtocells are nothing new, operating using existing over-the-air standards and user devices and offering services which are broadly familiar. As a result, it is expected that regulations in most countries, which already permit mobile services, will allow femtocell deployment with relatively little change. However, the deployment model for femtocells is sufficiently different to raise questions for some regulators and to require changes to some regulations to allow a broad range of business models to be supported. Sometimes simple uncertainty regarding the status of regulation can be enough to put operators off deploying a new technology, or discourage vendors from developing it, so it is vital to ensure that regulators understand the immediate nature of femtocell deployments, adapt regulation where required in a timely fashion and communicate the outcomes and implications clearly.

9.2 Regulatory Benefits of Femtocells

Before regulators can adapt or clarify regulations, they need to understand how femtocells will support their own regulatory goals. The specifics of such goals vary according to the regulator, but they would typically include a selection of the following:

- spectrum efficiency;
- economic efficiency;
- enabling competition;
- broadening access to services;
- enabling innovation;
- promoting competition;
- environmental goals.

It turns out that femtocells can help in meeting every one of these goals.

Femtocells: Opportunities and Challenges for Business and Technology Simon R. Saunders, Stuart Carlaw, Andrea Giustina, Ravi Raj Bhat, V. Srinivasa Rao and Rasa Siegberg © 2009 John Wiley & Sons, Ltd

9.3 Spectrum Efficiency

Spectrum efficiency is an important goal for many regulators. Spectrum enables a wide range of services and delivers economic and social benefits, but is usually thought of as being scarce, so the efficient use of spectrum serves to support several regulatory goals. As highlighted in Chapter 4, given the correct interference management techniques, femtocells are capable of operating in existing mobile spectrum, delivering very dense frequency reuse and consequent spectral efficiency. They also open up the use of higher frequencies, where there is still vacant spectrum but which is tricky to use with traditional macrocells due to the poor coverage and the resulting high costs. At low frequencies, the same interference management techniques also ensure the control of interference between adjacent femtocells. A single femtocell will cover an entire home at even exceedingly high frequencies, making femtocells capable of operation at almost any frequency where mobile devices can be made available. The increased spectrum efficiency of femtocells does not imply, however, that the overall requirement for mobile spectrum is necessarily decreased by the use of femtocells, since they may equally enable and stimulate an increase in demand for services which could not otherwise have been possible to deliver economically.

9.4 Economic Efficiency

Regulators have an interest in ensuring that mobile services are delivering at the lowest possible cost, making best use of the resources available, which is ultimately passed on to their national economy in terms of benefits generated for citizens/consumers and producers (in this case consisting of femtocell operators and vendors) alike. On the whole this is closely aligned with the desires of mobile operators to deliver services cheaply, but in a regulator's view the transfer of cost from an operator to the end-customer does not help, since the investment is still being made somewhere, which is ultimately passed on to the end-customer or to society as a cost. Femtocells instead deliver efficiency by reusing existing spectrum, making better use of existing investments in backhaul and reducing the total quantity of energy consumption.

9.5 Enabling Competition

Competition is seen by regulators as enabling efficient markets that drive down costs and inefficiencies in order to offer services to users at the lowest possible price compared with that which would be offered if there was no competition (monopoly) or little competition (oligopoly). It also encourages innovation, giving consumers greater choice and better products. Competition in mobile services is traditionally seen as challenging to promote, because of the high initial cost of infrastructure, and the limited supply of spectrum in which mass-produced phones are capable of operating. Femtocells can change this view, because service can potentially be offered to only those locations and customers that require them, and by enabling services to be offered by more operators than would otherwise be possible in a given spectrum band. By way of example, in 2006 the UK communications regulator, Ofcom, auctioned a 3.3 MHz block of spectrum, already supported by standard GSM 1800 MHz phones, to 12 operators, despite the block comprising only 15 standard GSM channels.[1] All 12 operators have access

---

[1] http://ofcom.org.uk/radiocomms/spectrumawards/completedawards/award_1781/.

to all 3.3 MHz over the whole of the UK. This is possible because the licences limit the power per base station to a very low level (equivalent to 23 dBm EIRP in one 200 kHz GSM channel). Although this was created with picocells in mind, this required active coordination between operators, while femtocells permit the end-users to deploy systems for themselves with a minimal likelihood of interference between operators. So the UK market now has a potentially much larger number of competing mobile operators than previously, enabled by femtocells.

9.6 Broadening Access to Services

Access to mobile services is increasingly seen as a necessity rather than a luxury. It enables users to conduct their work and leisure activities and drives economic growth (100). Yet for those who live in rural and sparsely populated areas, coverage is often poor and mobile operators have little incentive to build new sites which would not deliver a return for many years or possibly at all in their useful lifetime. Many regulators have imposed or are considering universal service obligations on operators for broadband services, which ensure that operators pay due attention to such areas. Femtocells allow operators to deliver mobile services even in isolated locations as long as backhaul is available. Also, access to computers is a limiting factor on access to the Internet, while mobile phones are available to a much bigger proportion of the population in many countries. Web-enabled phones permit personal access to the Internet on devices which are available to many. Of course, femtocells currently require access to fixed broadband lines, which may be a limiting factor in some rural environments. However, the potential for fixed wireless broadband access coupled to femtocells enables newer wireless technologies (e.g. WiMAX) to deliver connectivity to femtocells, which then permit legacy phones (e.g. GSM) to access services.

9.7 Enabling Innovation

Femtocells have the potential to act as a platform for new services and applications, enabling an ecosystem of developers to create applications for high bandwidth mobile services, unconstrained by the cost and bandwidth limitations of existing networks.

9.8 Environmental Goals

Femtocells provide a means to reduce the energy consumption and overall carbon footprint involved in delivery of a given service level. The power consumption of a stand-alone femtocell is currently around 10 W and is expected to reduce over time as increasing integration of key components takes place. The European Commission has published a code of conduct on broadband equipment (101) which includes targets for the power consumption of femtocells as shown in Table 9.1. The transmit power of femtocells naturally reduces when service

Table 9.1 European Commission targets for the power consumption of femtocells

	2009/10		2011	
Introduction of new model	Low-power state	On state	Low-power state	On state
Target maximum power consumption	9.0 W	9.0 W	7.0 W	8.0 W

is not being delivered and there is additional scope for intelligent power-saving measures. Femtocells integrated with broadband routers are expected to consume even less power. The large quantity of macrocells required to deliver a given level of coverage improvement will consume significantly more power. A full environmental assessment requires consideration of the complete lifecycle for femtocells, including the production and disposal costs of the femtocell. Nevertheless it is expected that significant benefits can be realised.

9.9 Spectrum Licensing Issues

Since femtocells operate over the air to produce identical signals to those produced by existing standards-based base stations, in most locations existing spectrum licensing regimes already allow the operation of femtocells. The main purpose of spectrum licensing is to avoid harmful interference between licensees, and the low power and interference management capabilities of femtocells mean that there is virtually no likelihood of interference between operators. Additionally the regulator needs assurance that the femtocell can only be operated by the licensed operator and that the user cannot tamper with the femtocell and thereby force it to operate illegally.

However, in some countries spectrum licenses include requirements for the licensing and registration of individual base station equipment which may present a significant barrier to efficient femtocell deployment. Such requirements can include:

- The need for base stations to be installed by specially qualified personnel. For example, until late 2008 the regulations in Japan stipulated that only licensees with specialist skills can operate radio systems. In the worst-case interpretation of this regulation, a base station engineer would have to be present whenever it was required to switch on a femtocell.
- The payment of fees for each base station commissioned. For example, it has been reported that the regulator in Hungary requires a payment of €1000 for each base station.
- The provision and maintenance of records of base station locations. While operators will typically require knowledge of femtocell locations to ensure that spectrum is only used where it is licensed and that users are correctly provided with services, the practicalities of providing records suitable for regulators, yet applicable to potentially millions of femtocells, make this difficult. In some cases the motivation for these records is to provide information to the general public with relevance to concerns regarding health and safety. Adding millions of low-power cells may reduce the value of these databases and create unnecessary and ill-founded concern.

These regulations may be entirely reasonable when used for their original purpose of licensing macrocell base stations, but can create a significant barrier to femtocell adoption. Such regulations will therefore need modification in advance of large-scale femtocell deployment.

It has been suggested that, given femtocells' low tendency to cause interference, they could be classed as license-exempt (unlicensed) devices. For example, transmission from mobile terminals is typically license-exempt because the mobiles transmit only under the command of operator-controlled base stations. However, femtocells are directly part of the operator network, and are network elements in the standards, so it is more appropriate to treat them under existing base station licensing frameworks. It is, however, entirely possible that femtocell-based technology might be used in the future for operation in license-exempt spectrum, though in this case they would not benefit from the large existing user base with handsets and would indeed not formally be femtocells at all according to the definition in Section 1.4.

9.10 Location

There are several motivations for femtocells to be capable of reporting their location:

- To ensure that femtocells only transmit where they are licensed to operate. A user may move the femtocell from their registered address and the femtocell has to be capable of detecting this and reporting it to the operator for checking before transmission is enabled. A particular example is in the United States, where operators typically have very different spectrum holdings in different markets.
- To deliver the call location when the femtocell is delivering an emergency call or is subject to lawful intercept.
- To enable efficient interference management, for example ensuring that the femtocell does not reuse scrambling codes used or planned to be used by nearby macrocells.

There are numerous methods for delivering this location capability as described in Section 10.3. Given all of these techniques, it should be possible for femtocells to report their location with a high likelihood and good precision in the majority of cases. However, there may still be occasional cases when none of the available techniques yields adequate location information. In these rare circumstances the femtocell may have to deny service.

9.11 Authentication

Femtocell manufacturers must ensure that only approved femtocells operate on a given operator's network, and that each femtocell is permitted to access service from the chosen operator. This implies that the femtocell and operator network must be mutually authenticated in a fashion that is secure and tamper proof in order to assure regulators that femtocell technology is being used appropriately. These issues are discussed in more detail in Chapter 7.

9.12 Emergency Calls

In most jurisdictions it is required that emergency calls are handled by mobile systems, and that the calls are delivered together with information on the location from which the call is being made. For example in the United States, the Phase II E911 emergency calling rules laid down by the Federal Communications Commission require wireless service providers to provide the latitude and longitude of the caller to within 50 or 300 metres depending on the type of technology used (102). As discussed earlier, there are several ways of delivering location information. In some cases, however, it may be easiest to deliver location information specifically for the purpose of emergency calls by simply ensuring that emergency calls are carried on the macrocell network. Although this may not be appropriate where improved coverage is the main motivation for the femtocell, in fact 2G coverage may often be available where 3G coverage is not, so this may well be appropriate in many cases.

In some environments, operators must carry emergency calls originating from phones of customers from different operators' networks, and sometimes even when no SIM is available. These regulations may not necessarily apply to femtocells in all locations.

It is important to recognise that femtocells are expected to enhance the access to emergency service compared with the case when femtocells are not present. However, it is necessary for customers to understand any limitations. For example, if the power supply to a femtocell is switched off by the user, they must appreciate that they will not have access to service and will require an alternative means of access to emergency calls. This is a similar situation to the use of existing cordless phones (e.g. DECT-based systems), but requires that proper labelling and information is made available as part of femtocell offers.

9.13 Lawful Interception and Local IP Access

As described in Chapter 5, the concept of local IP access using femtocells offers an efficient manner of transporting mobile data traffic without having to upgrade the capacity of the mobile core network and without losing control over the use of licensed spectrum. Traffic is routed away from the mobile network, either at the femtocell and home router in the case of traffic which is moving around the home LAN or Internet; or at the femto gateway in the case of other traffic destined for the Internet.

However, while such an arrangement is cost efficient and avoids unnecessary 'tromboning' of traffic to and from the mobile core, it does raise some regulatory issues associated with lawful interception and emergency call handling.

Lawful interception (LI) is the means by which law enforcement agencies obtain network traffic data, typically in real time, for the purpose of analysis or evidence. Such interception is a legal requirement on networks in most parts of the world. For example, in the European Union a European Council resolution mandates such measures on all EU member states (103). In the United States the Communications Assistance for Law Enforcement Act (104) provides the statutory framework for the duty of network operators to deliver forensic evidence for law enforcement. More widely there is the global Convention on Cybercrime which covers many international signatories (105).

A reference architecture specifying how the network operators and law enforcement agencies interact for the purposes of LI is specified by ETSI (106). In conventional mobile networks, the LI capability is implemented in the core network, since all traffic, both voice and data, has to pass through the core and associated gateways. It is also clear that the responsibility for delivering LI rests with the mobile operator, whose network is in use for the whole of the service.

The responsibility is less clear in the case of femtocells implementing local IP access. Given that the end-customer may be contracting with both the mobile operator and a fixed operator, which of these has the responsibility? In some countries, the use of licensed spectrum implies that the mobile operator is responsible. However, apart from the short mobile link, the femtocell traffic is essentially indistinguishable from other traffic (e.g. VoIP traffic, including VoIP traffic carried over Wi-Fi) which is carried by the fixed operator, who then carries the responsibility. Why should femtocell traffic be treated differently from this? Is traffic that flows only around the user's home subject to the same LI requirements? The answers to these questions may well differ amongst regulators in different locations and must be taken into account by vendors selling into those markets.

The mobile operator could discharge the responsibility by rerouting the traffic into the core when LI is required. However, this will change the traffic flowing from the femtocell and will be noticeable by a knowledgeable user, so this may not be an acceptable solution. In this case the opportunity for local IP access, at least for home traffic, may be lost. Clarity on these regulations – and consistency with other services – will be important for the industry to provide compliant but efficient systems.

9.14 Backhaul Challenges

Delivering a femtocell service to an end-customer involves a combination of fixed and mobile network assets, which may confer a distinct advantage on a converged operator who can control and operate both the mobile access over licensed spectrum and the backhaul over their fixed network in an efficient, joined-up fashion. However, it is entirely feasible to operate a femtocell service where the two elements of the network are delivered by different operators and indeed where there is no direct commercial relationship between those operators (see Section 11.2 for a discussion of the various potential relationships between operators delivering elements of a femtocell service). This scenario does create commercial challenges, however, where the fixed operator could take the view that they are carrying the femtocell traffic without any commercial benefit. Their customers could equally take the view that they are paying for their broadband service and expect to be able to run applications over that broadband, including femtocells, without hindrance. Regulations may affect the situation.

In some regulatory environments, explicit 'net neutrality' provisions are in place, preventing – or at least discouraging – broadband operators from treating customer traffic associated with particular applications differently. Such regulations would assist in ensuring that femtocell traffic at least has an equal opportunity to be carried with equal priority amongst other traffic. It does not, however, ensure that real-time applications such as voice will get sufficient priority to be delivered with good quality (98). Also, such regulations are relatively rare, with most jurisdictions deciding that the Internet has grown successfully without such regulations, and that it is better to act to ensure that the general regulatory environment allows for sufficient competition amongst broadband operators to ensure that innovations such as femtocells can flourish on appropriate commercial terms.

Thus, fixed operators may well wish to enter into commercial arrangements for the delivery of femtocell services with mobile operators. This will provide a competitive advantage for their 'femto-friendly' broadband service over other operators, and will give them the opportunity to provide services to a potentially very large customer base, in a situation where it is likely that the mobile operator will contract with only a limited number of Internet service providers.

9.15 Mobile Termination Rates

Termination rates are the charges that one network operator (mobile or fixed) charges another operator for terminating voice calls on its network. Since such termination fees are not subject to competition as far as the calling party is concerned, termination rates are widely regulated. Although the specific approach to regulation varies around the world, they are typically based on international benchmarks or on models which attempt to determine the incremental cost of carrying an additional call by the terminating operator. Femtocells are not expected to affect these arrangements directly at first, but the differing cost base for femtocells compared with conventional mobile networks could over time cause femtocells to be a relevant consideration for cost modelling. However, the cost benefits of femtocells are probably most relevant to carrying data traffic, where termination rates are not regulated.

9.16 Competition Concerns

Although femtocells are expected in the main to enhance competition in the mobile sector, there are some potential concerns that have been expressed by regulators. The delivery of femtocell 'bundles', which include the femtocell service along with broadband and wide-area mobile access and other services, may make it harder for customers to switch providers. Similarly, femtocell services might not be available from mobile virtual network operators (MVNOs), if these are not covered by the agreements between the MVNO and the operator whose licensed spectrum is used. However, these issues are generic to any bundled service or other service offered by a given operator. The operation of femtocells will not prevent customers from continuing to adopt services from operators who do not offer femtocells if they consider the service acceptable. Likewise, operators with any spectrum band can offer femtocells on a similar basis and cost, so femtocells may actually reduce any competition concerns arising from different spectrum holdings. On the whole, therefore, these issues should not raise major regulatory concerns and are 'business as usual' in an already competitive industry.

9.17 Equipment Approvals

To enable sale in many markets, femtocells must undergo appropriate approval procedures, applicable to their status as both consumer electronics devices and as intentional radio transmitters. For example, within the European Union equipment manufacturers must demonstrate compliance with the essential requirements of the Radio and Telecommunications Terminal Equipment (R&TTE) directive (107). This includes aspects such as:

- protection of the health and safety of the user;
- ensuring compliance with electromagnetic compatibility requirements;
- effective use of radio spectrum so as to avoid harmful interference.

Meeting such requirements often involves demonstrating compliance with relevant harmonised standards and may require testing and declaration by the manufacturer or testing by approved test houses depending on the locality. The specific requirements vary substantially around the world and manufacturers should take care to ensure compliance with local regulations in their target markets.

9.18 Examples of Femtocell Regulations

There has been progress on femtocell regulatory issues in many parts of the world. Some examples are given here.

In Japan, noting that there were several aspects of the existing regulations which were not entirely aligned to femtocells, the Japanese regulators conducted a series of consultations during 2008, and announced the outcome in December 2008. In summary, the outcome is:

- A femtocell is defined for regulatory purposes as a special form of base station, meeting relevant standards but having an output power of no more than 20 mW into a 2 dBi gain antenna. It should be difficult for the user to open, and be capable of halting transmissions in the case of a fault being detected. Transmissions should also be prevented if connectivity to the operator over the broadband line is lost.

- Femtocells can be powered up by end-users, in contrast to the previous regulations where this was only permitted for trained personnel.
- The mobile operator (rather than the broadband operator) is responsible for the femtocell service, but the broadband line can be used for femtocell backhaul.
- No back-up battery is required.
- The operator should provide adequate end-to-end speech quality and data connectivity via the femtocell. To facilitate this, appropriate discussions between the mobile and broadband operators are a requirement of the regulations.
- Users should be properly informed of the capabilities and limitations of the femtocell service.

In Europe, the body responsible for developing measures to implement common radio spectrum policy issues across the 27 member states of the European Union is the Radio Spectrum Committee (RSC). It has the ability to create decisions which are binding on member states under European law. In 2008 the RSC considered spectrum issues on femtocells. It decided that, in view of the control which operators can exert over femtocells as part of their existing network, femtocells could operate under the existing spectrum-licensing regimes of member states and there was no current need for the RSC to take action. They also noted that the increased spectrum efficiency available from femtocells was a positive development (108):

> Noting that femtocells operate as part of the operator's existing network (using the same frequencies) and that the operator remains in control of the femtocell at all times, it is reasonable therefore to assume that femtocells will comply with the existing technical licensing conditions in each specific case.
>
> Consequently, any intention of the Radio Spectrum Committee to engage in the development of any specific regulation concerning femtocells should be carefully justified on the basis of a clear added value. Nevertheless, the proliferation of femtocells is supported in the context of more efficient use of spectrum.

The UK communications regulator, Ofcom, published a review of the mobile sector in August 2008 (10). While this did not contain any specific policy proposals relating to femtocells, there were numerous references to them, recognising their significance and potential impact:

- It noted that technology such as femtocells could actually reduce the trend for reduction in fixed-line services by cementing household reliance on fixed line, albeit to deliver mobile service rather than conventional fixed telephony.
- It recognised that femtocells are part of a new wave of developments which could lead to networks being neither fully fixed nor fully mobile:

> 'they form part of a vanguard of a long-promised technology that has the potential to enable new forms of competition across communications networks: fixed-mobile convergence'.
> 'Femtocells are one of a number of technologies (alongside VoIP and UMA) that substantially alter the way that voice calls are routed across mobile and fixed networks'.
> 'Femtocells enable operators to target in-home usage more accurately and at lower costs than conventional delivery'.

- It noted that femtocells may lead to a need to question traditional regulatory assumptions which regulated fixed and mobile entirely separately:

'For example, would terminating a call over a fixed broadband backhaul connection to an in-home femtocell be considered fixed or mobile voice call termination?'.

9.19 Conclusions

In the main, femtocells are capable of operating under existing regulatory regimes with little need for change. However, there are some marginal issues where regulatory changes will be of assistance to the industry and regulators alike to realise their regulatory benefits, and there are other areas where some interesting new issues are created, primarily resulting from the converged nature of femtocells, where there may be some difficulty in classifying femtocell traffic as either fully mobile or fully fixed. The industry and the regulators need to address these issues promptly and preferably consistently to ensure that early clarity is available to avoid delaying or impeding femtocell deployments.

10

Femtocell Implementation Considerations

Simon Saunders

10.1 Introduction

The design and implementation of femtocells is non-trivial. There is a wide range of engineering and commercial issues, involving many engineering disciplines, which have to be addressed to meet the requirements described in previous chapters. This involves a considerable focus on the underlying hardware and software architecture and cost. The major factors are discussed in this chapter. Overriding all of these is the need to deliver the full set of desired features and the levels of performance required by standards and operators, while ensuring that the cost is sufficiently low to support a strong business case. In all cases this will be supported by a high volume market with significant levels of commonality of parts and functionality, at least as far as the major components and undifferentiated software elements are concerned.

The main challenges we address here are as follows:

- signal processing;
- location;
- frequency and timing control;
- protocol implementation;
- RF implementation.

We also examine the femtocell hardware architecture and the associated cost challenges. Finally the opportunities and challenges for optimising mobile devices for enhanced femtocell operation are highlighted.

Femtocells: Opportunities and Challenges for Business and Technology   Simon R. Saunders, Stuart Carlaw, Andrea Giustina, Ravi Raj Bhat, V. Srinivasa Rao and Rasa Siegberg   © 2009 John Wiley & Sons, Ltd

10.2 Signal Processing

The emergence of sufficiently cheap signal processing to deliver femtocells is one of the critical factors for their implementation, as discussed in Section 2.5.7. A 3G femtocell needs several hundred MIPS – millions of instructions per second – of processing capability (33). While the trends in signal processing are certainly towards rapid increases in the processing capacity to be delivered at a given cost point, it is necessary to accelerate this for femtocell applications. This requires some careful choices amongst technologies, including DSPs, FPGAs, specialised wireless signal processors, ASICs and systems-on-chips. The appropriate selection will depend on the timescale for implementation, the market size and the mixture of features required for the application. There is also a need to adopt consumer electronics approaches rather than those that have traditionally been employed in wireless infrastructure.

10.3 Location

To support both operator and regulatory requirements, femtocells must be capable of reporting their location back to the associated management system before they start to transmit and then regularly thereafter, certainly whenever the location changes appreciably. There are numerous methods for delivering this location capability:

- *Self-reporting by femtocell customer*: typically a femtocell will be registered to a particular customer address, and this gives a first indication of the expected location. In some instances this may be sufficient if the customer is given and accepts clear information as to the consequences of providing incorrect information. In other cases this will be insufficient, but it gives a useful starting point for comparison with reports delivered by the femtocell via other means.
- *GPS*: although for navigation applications GPS delivers very poor indoor availability, given long enough time to process signals, specialised GPS receivers can detect their location at exceedingly low signal strengths, so that GPS becomes viable in many environments. This may be inconvenient for users, however, who may be requested to place femtocells close to windows and/or deploy an external GPS antenna.
- *Macrocell detection*: by reporting the surrounding macrocell identities to the management system, the femtocell location can be determined to a good degree of precision, suitable for most regulatory purposes. Although the use of femtocells to improve coverage may suggest that macrocell coverage may not be detectable, in fact the detection of a macrocell for simply decoding its cell identity can be achieved at considerably lower signal strength than that required for delivery of acceptable service, increasing the number of environments where this is a useful approach. Additionally many 3G femtocells will detect 2G cells, which typically deliver better coverage. Similarly it has been proposed to use television transmitters as reference sources for femtocell location.
- *Broadband port identity*: given that a femtocell achieves its backhaul via the fixed broadband network, the broadband operator can locate a femtocell physically via knowledge of the local exchange or network port from which service is accessed. This approach produces a very good degree of confidence in the location, and is directly available in the case of an integrated operator who is delivering both the fixed and wireless portions of the service. In the case

of separate operators, the approach will require either a commercial agreement between the relevant operators or a regulatory compulsion on the fixed operator. Such regulatory arrangements are already being proposed in many parts of the world to enable support for emergency call location by VoIP services, with all broadband operators contributing location details of broadband subscribers to a central database.

Given all of these techniques, it should be possible for femtocells to report their location with a high likelihood and good precision in the majority of cases.

10.4 Frequency and Timing Control

The needs for frequency and timing control differ significantly between air interface standards. WCDMA systems typically require only frequency control and are unsynchronised at the level of base station timing. The requirements are more stringent in CDMA, where the phase of transmission timing also has to be carefully controlled. In TDD systems such as TD-LTE and WiMAX it is also important that the split between uplink and downlink timing is appropriately synchronised.

One option to meet many requirements is to use GPS, providing both timing and location data in a single solution. GPS can, perhaps surprisingly, deliver quite reasonable indoor coverage by using specialised receivers which integrate the GPS signal over an extended time period to extract it from the noise, taking advantage of the fact that this is intrinsically a stationary application. GPS is certainly a solution in use in some commercial femtocells. However, this comes at a relatively significant cost and requires an extended acquisition time in locations with very low signal strengths, which is apparent to the user in the initial set-up time between powering on the femtocell and achieving service. It may also require users to place the GPS antenna close to a window, which may be unacceptable to some.

The macrocell network can be used to acquire both frequency and timing information. This clearly depends on there being sufficient macrocell signal available, although the required signal strength is much lower for this purpose than for obtaining a reliable service. It cannot, however, be relied on in all cases.

Relaxations in the required frequency accuracy for femtocells have helped in reducing the precision of the basic frequency reference oscillator within the femtocell. For example, in WCDMA, femtocell basic accuracy only has to be 250 parts per billion (ppb), compared with 100 ppb for local area base stations and 50 ppb for macrocells. This level of accuracy is still challenging to achieve at an appropriate price. A simple crystal oscillator, while relatively low cost, does not deliver the required stability and some temperature control and/or compensation is necessary. A temperature-controlled crystal oscillator (TCXO) delivers typically 10 times better frequency stability, but may still not be sufficient in its basic form. Greater stability, around 1000 times better than a basic crystal oscillator, may be achieved via an oven-controlled crystal oscillator (OCXO) but at a significant cost and power consumption. One approach is to use a TCXO but with a network-based approach to periodically recalibrating the frequency.

This is challenging in itself, due to the jitter introduced by the backhaul network, but solutions such as the precision time protocol (PTP) within the IEEE 1588 synchronisation standard (109) or the IETF network time protocol (NTP) (110) may be suitable.

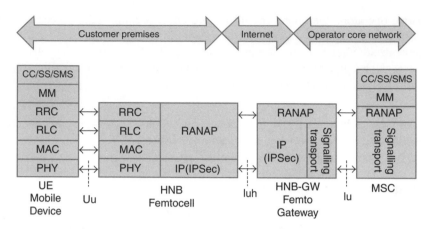

Figure 10.1   Protocol stack for a WCDMA femtocell network

10.5 Protocol Implementation

The femtocell requires a significant amount of protocol software in order to operate. The functional split agreed in 3GPP for WCDMA femtocells, for example, requires that the femtocell incorporates all of the functions of a conventional base station but with the addition of most of the functionality of an RNC. The relevant protocol elements are divided between the femtocell (HNB), gateway (HNB-GW) and switch (MSC) as shown in Figure 10.1. In addition, the femtocell will usually incorporate part of the functionality of a UE in order to determine the surrounding macrocell environment, plus algorithms for interference management, plus the device and security management systems. Overall there is a significant software development and maintenance effort which must be achieved efficiently in the femtocell. It must be compliant with the standard and also demonstrated to interwork with other network elements.

10.6 RF Implementation

In addition to the timing and synchronisation, the femtocell also requires a complete RF subsystem. Although this is a low-power device, the designer faces a number of implementation challenges:

- Price requirement comparable to that for mobile devices, but with lower volumes and different RF requirements.
- The need for a receiver operating in the downlink band as well as one for the uplink.
- The need to avoid suffering and creating interference via blocking and spurious emissions to and from nearby devices such as integrated Wi-Fi, nearby DECT etc.
- The need to support multiple frequency bands in different product variants.

Implementation of femtocell RF elements draws heavily on devices intended for handsets, but is likely to evolve rapidly towards the use of custom RF components to support the particular needs.

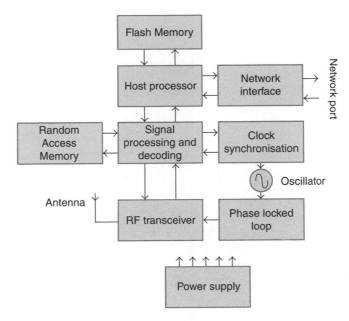

Figure 10.2   Example femtocell hardware architecture

10.7 System Design and Cost

Figure 10.2 shows a high-level hardware architecture for a stand-alone femtocell, identifying the major elements. Figure 10.3 shows how these elements can be realised in hardware in one example reference design.

All of these elements contribute to the unit cost of the femtocell. Figure 10.4 shows an indicative breakdown of the costs associated with the hardware elements of a femtocell and how these might vary with time and with the approach taken to implement the necessary baseband signal processing. In this figure, 'standard processing' indicates that signal processing is conducted using general-purpose DSP/FPGA hardware, while 'specialised processing' indicates the use of processing elements targeted specifically at the needs of wireless baseband processing, which can reduce the associated costs. The 'future' cases indicate the potential for cost reduction within a few years, given healthy take-up of femtocells. Note that, as well as the signal processing, the processing needs for encryption may also be significant as discussed in Chapter 7.

It is expected that increasing levels of integration, higher volumes and processing, which is increasingly specialised for femtocells, will progressively reduce the overall hardware costs. This is also supported by the general trends for cost reduction in general-purpose components such as memories, leading to significant opportunity to reduce the cost substantially as the take-up increases. There is a critical trade-off in the design between the need to maintain flexibility, permitting the evolving standards to be reflected, addressing a wide range of operator and customer needs for new features, and the desire to reduce costs rapidly. Nevertheless the opportunity for cost reduction is substantial, given general trends in consumer electronics. For comparison, early Wi-Fi access points cost many thousands of dollars.

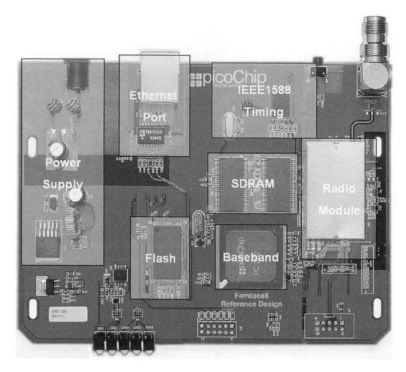

Figure 10.3 An example femtocell reference design, showing major components. *Reproduced by permission of picoChip Designs Ltd*

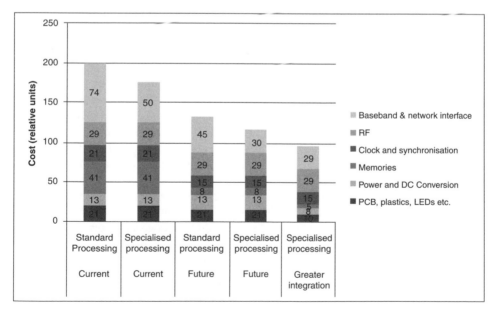

Figure 10.4 Example bill of materials cost analysis for femtocells

However, some of the elements, such as the power supply and the casing and displays of the femtocell show relatively little opportunity for cost reduction. In these cases, however, there is an opportunity for sharing the costs by integrating the femtocell functionality with home gateways which already comprise several functions such as DSL or cable modem, router, Wi-Fi access point etc. A similar opportunity is available from integrating femtocells with a wide variety of other possible consumer devices, such as set-top boxes, home media servers, home computers, games consoles and even digital photo frames.

Another opportunity for cost reduction is to ensure the standardisation of the functionality and interfaces between the various hardware elements, increasing the economies of scale for manufacturers of the elements. For example, there is potential to harmonise the interfaces between the transceiver and the baseband signal processing (modem) as well as that between the modem and the remainder of the processing. There is also scope to harmonise the module interface for a femtocell intended for integration into another device. All of these hardware interfaces are being addressed by the Femto Forum.

It is important to point out that the hardware is not the only cost element for femtocells. A very substantial quantity of software must be written and maintained, including protocol stacks, management software, application software, security features, the algorithms for autonomous provisioning and optimisation and so on. This software also represents a substantial opportunity for developers to differentiate their products and for flexibility in the features offered by operators to attract their customers. The software resides not only on the femtocell itself, but also on the gateway and management systems. The developers of this software need to affray the development and maintenance costs associated with this software over the volume of femtocells delivered and this must be carefully considered in an overall analysis of femtocell costs.

10.8 Mobile Device Challenges and Opportunities

A central design requirement for femtocells is the need to support existing mobile devices with absolutely no changes, ensuring a positive experience for the vast population of existing users. However, there are opportunities for mobile devices to be further optimised for femtocell operation. These opportunities include:

- Support for specific femtocell-related mobility management functions, enabling further enhancement of battery life by reducing the quantity of measurements that have to be performed by the device in support of handover decisions. The overall battery life is substantially greater for even legacy devices on femtocells for a given usage pattern due to the reduced transmit power, but it is likely that users will be encouraged by the services and performance of the femtocell to use their devices more, which may offset some of the gains, so additional battery life savings can be valuable.
- Support for closed user group functionality, which can allow users to select individual femtocells or groups of femtocells, and for operators to control access to femtocells. This enables a wide range of potential new business models and applications, but requires new devices. Such features are enabled in standards, via the CSG functionalities in 3GPP and in the WiMAX Forum and the comparable PUZL capabilities in 3GPP2. There are questions for manufacturers as to how to present these similar but different capabilities in a consistent manner for multi-mode handsets supporting multiple standards.

- In a related point, the 'connection manager' approach taken in the mobile device to select amongst air interfaces according to the available networks, billing, applications and operator relationships may differ somewhat for femtocell-optimised operation.
- Femtocell-specific service and application capabilities (such as those described in Section 11.5) may benefit from specific functions and user interfaces in mobile devices.

Thus the potential optimisations to mobile devices for specific femtocell awareness can span several different elements of the device, including the ergonomics and form factor, the protocol stack, the operating system, mobility management applications and user interface. These changes could take two to three years from inception to reach commercially sold devices and should be considered by device vendors and operators at an early stage to maximise the associated opportunities. See (111) for further discussion of these issues.

10.9 Conclusions

The successful development and manufacture of femtocells requires the coordinated operation of a range of disciplines, including digital and analogue hardware and modem, protocol and application software development. These have to be brought together to deliver standards-compliant devices which also enable a differentiated feature set, all at a cost consistent with an attractive operator business case. The overriding enablers for this are complete and timely standards and the reusability of subsystems for different applications, all enabling and being enabled by increased volumes.

11

Business and Service Options for Femtocells

Simon Saunders and Stuart Carlaw (Section 11.7)

11.1 Introduction

The advent of femtocells enables more than simply access to the same services as in the macrocell network, supplied more cheaply and efficiently. It also permits new ways by which mobile-based services can be delivered, via different constellations of operators and vendors. It further enables the services delivered to extend beyond those which could be delivered by the macrocell network alone. This chapter explores some of these new possibilities. While many are at their early stages and are not yet seen in the marketplace, all are based on a desire to explore the boundaries of femtocell enablement and the need to ensure that femtocell equipment, software and standards developers are thinking through the full range of potential requirements.

11.2 Ways of being a Femtocell Operator

Given that femtocells necessarily operate in licensed spectrum, it should be clear that mobile operators – or at least those holding licences for spectrum permitting mobile use – form an essential component of any delivery of a femtocell service.

Yet that requirement still allows a wide variety of possible delivery approaches. These range from fully vertically integrated approaches, where all functions are fulfilled by a single operator, through to cases where all of the elements of service are 'unbundled' and the spectrum is held by a spectrum broker.

These distinct cases are characterised by the entities who own, or at least control, the various elements of end-to-end femtocell service delivery, as illustrated in Figure 11.1.

In a conventional service delivery model, the mobile operator owns the spectrum, gateway and management system. Their service will be carried over the Internet via an Internet service

Femtocells: Opportunities and Challenges for Business and Technology   Simon R. Saunders, Stuart Carlaw, Andrea Giustina,
Ravi Raj Bhat, V. Srinivasa Rao and Rasa Siegberg   © 2009 John Wiley & Sons, Ltd

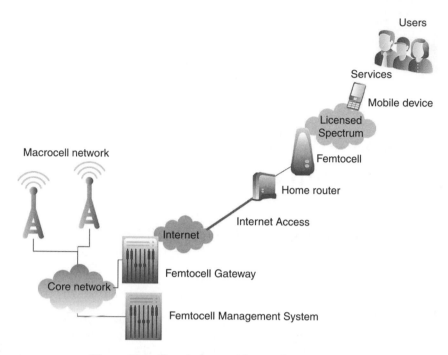

Figure 11.1   Key elements of femtocell service delivery

provider who delivers the broadband service to the user. The user's macrocell service is provided by the same operator. The user therefore contracts with both the mobile operator and the ISP for their service. There is no commercial relationship between the mobile operator and the ISP. This model is entirely viable and likely for deployments in many parts of the world, enabled by the efficient interfaces and protocols for femtocell backhaul over standard broadband described in Chapter 5.

In a more integrated case, the mobile operator and the ISP would be the same operator. Femtocells are one of many forces that are leading to the proliferation of such integrated operators. Such operators include both mobile operators, who have started offering fixed-line Internet service (often via the mechanism of local loop unbundling), and also fixed-line operators, who have either acquired mobile spectrum themselves or have some form of legitimate access to it. This gives the operator the opportunity to control the broadband service quality, but more importantly, it gives them the ability to deliver integrated service packages to the end-user, who only has to engage with a single operator and can gain the benefit of services which are more uniformly delivered and attractively priced, while the operator gains increased customer loyalty and an opportunity to provide additional value-added services. Even in this highly integrated case, the services accessed by the customer include both operator-delivered services and Internet service provided by third parties where the operator does not participate, other than in the enablement of access.

In some cases, full integration of a fixed and mobile operation may be neither desirable nor permitted by local regulations. In this case a mobile operator could still achieve the same service possibilities as in the integrated case by contracting at wholesale level with a fixed-line operator. The fixed operator provides wholesale broadband service to the mobile operator and may also be responsible for operating the femtocell management system, for reporting faults and status to the mobile operator via links between the relevant management systems and for shaping traffic across a combination of the fixed network and the femtocells in an efficient manner. The service offered to the end-customer is indistinguishable from the previous case. This approach plays to the strengths of the fixed operator, who is used to managing networks based on the Broadband Forum or DOCSIS approaches and containing millions of nodes, which is new territory for the mobile operator. In this case it is important for the relationship between the fixed and mobile operators to be structured around an appropriate service-level agreement, which specifies all the important dimensions of service quality and reliability between the operators as well as the quality experienced by the end-customer.

The converse of the previous model is where the fixed operator contracts with the mobile operator for the use of their spectrum and core/macrocell network, acting as a mobile virtual network operator (MVNO). The user can again buy full service bundles comprising both fixed and mobile services, but this time they contract only with the fixed operator. The service in the macrocell environment may be handled in the same way, with the fixed operator appearing as the service provider as far as the customer is concerned, or may be provided on a national roaming basis with one or more mobile operators. In some markets, consumers may be more comfortable with their home service being provided by a fixed-line operator, in which case this model could lead to greater take-up for all concerned.

In an extreme case, we could imagine a situation where many of the elements are provided by different entities. The spectrum is provided by a spectrum management organisation (SMO), who leases the spectrum to a variety of players, possibly in different geographical 'lots' in different parts of the same country. The macrocell network is operated by a managed service provider, potentially a large-scale network equipment vendor. The fixed network is physically operated by a traditional fixed telecommunications operator, while Internet access is provided on an unbundled local loop by a pure ISP. The femtocell is bought as a retail device by the end-user, but the user purchases their femtocell service from a large grocery retailer who operates a special femtocell MVNO service to allow this. When outside their home and travelling between locations, the user can access wide-area mobile services from a conventional wide-area mobile operator. In their workplace they can access a company-owned femtocell service, based on spectrum rented from the same spectrum management organisation. In cafes and airports the user accesses femtocell service on a public femtocell 'hotspot' service, again using spectrum rented from the SMO, but this time by an organisation best known for its Wi-Fi hotspots.

The reader can doubtless imagine further possibilities. While some of these examples – and particularly the last one – may seem fanciful and may well not be permitted by regulations in a particular territory, the point is that femtocells may open new service possibilities and business models, playing to the strengths of different entities, which may ultimately lead to a more efficient and innovative service with greater take-up and benefits for the end-user and better business for all concerned. Indeed, developments such as femtocells – amongst many others – may lead to traditional distinctions between 'fixed' and 'mobile' operators as such becoming increasingly irrelevant. Instead there is simply a wide set of operators of various services, each with their own particular assets and skills, who can engage with other operators with complementary capabilities to deliver the full end-to-end service to the delight of their customers.

Operators and developers of femtocells will need to consider the impact of all these service possibilities to access the widest possible market for their goods and services.

11.3 Femtocells for Fixed-Line Operators

As illustrated in the previous section, fixed-line (wireline) operators play an important role in the delivery of femtocell services and should consider their response accordingly. Here we briefly consider the role of femtocells from their perspective.

Fixed-line operators in many parts of the world face similar challenges, such as:

- Declining customer base and declining revenue from traditional services, especially voice.
- Fierce competition from large numbers of ISPs as well as from mobile broadband.
- Increasing use of VoIP services delivered by operators with a much lower cost base.
- Increasing tendency to deliver only 'pipes' for data and voice, with limited opportunity for new applications.
- Increasing fixed-mobile substitution for both voice and broadband data.

Finding some means to access the continued growth of mobile services, making the most of their assets, is a priority for most fixed operators. Femtocells can give them access to such opportunities, including:

- Building loyalty via quad-play service bundles.
- Accessing the growth of mobile broadband services.
- Enhancing opportunities for customer ownership relative to conventional MVNO arrangements.
- Differentiating their broadband offering against competitors via 'femto-friendly broadband'.
- Delivering a managed service to mobile operators, using existing skills and assets particularly based on TR-069 and DOCSIS management approaches and systems.
- Accessing an immense installed base of mobile users.
- Making the trend to fixed–mobile substitution less relevant: femtocells need fixed broadband!

What do fixed-line operators have which may be attractive to mobile operators for cooperating on a femtocell service?

- First and foremost – fixed lines! An *almost* essential component of femtocell services.
- Consumer market research studies for femtocells show consumers sometimes see the broadband operator as the natural supplier of femtocells.
- Femtocells drive further broadband penetration – even to customers with no interest in computers – thus increasing their customer base and revenue.
- Femtocell services will drive bandwidth demand and hence the potential to sell higher bandwidth, particularly with differentiated QoS, at premium price.
- Increase the revenue associated with supply and management of femto-integrated home gateways.

Overall, femtocells deliver fixed operators a number of potential benefits which can be accessed without the need to acquire spectrum. This does not necessarily come at the expense of the mobile operator, who benefits from reduced operational complexity, more credible service for the end user, reduced capital and operational costs, and a more rapid way to achieve a presence within the home.

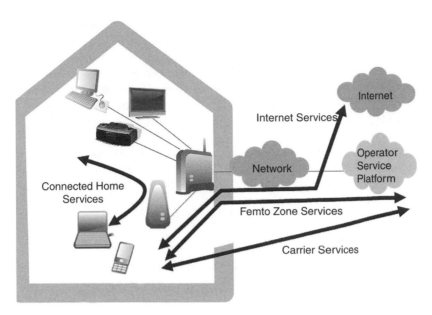

Figure 11.2   Four types of femtocell service. *Reproduced by permission of Femto Forum Ltd*

11.4 Types of Femtocell Service

Femtocells enable a wide range of services, some of which are existing services delivered in a more efficient manner, and others that are entirely new. To assist in distinguishing these services, the Femto Forum has proposed a separation of services into four categories, described as follows and illustrated in Figure 11.2.

- **Carrier services** – the conventional wide area service set offered by an operator to customers. Service examples are mobile voice calls, SMS, push-to-talk, etc.
- **Internet services** – services that are based on third-party platforms, accessible by any Internet access method (e.g. broadband or mobile broadband). Service examples are Web browsing, e-mail, ftp, mobile TV etc.
- **Femtozone services** – services where the femtocell enables or assists in the delivery of the service via service information from or service execution at the femtocell. Service examples are 'virtual home phone', home presence, etc.
- **Connected home services** – services delivered to the mobile device from the home network while a subscriber is in the home environment. Service examples are automated music synchronisation, use of the mobile as a home automation remote control, mobile as a media player to listen to home music collection or watch home videos, etc.

Note that all four service classes are only possible for a user if they are permitted by the operator, by virtue of the operator's management control of the femtocell and the regulatory constraints of licensed spectrum.

Operators may offer only carrier and Internet services, or a combination of all four classes, depending on their motivations for delivering femtocells and local market and regulatory conditions. However, while carrier and Internet services have been the starting point for much of the industry due to the increased efficiency in delivering services for the operator, the femtozone and connected home services offer particular promise for driving additional revenue and for helping the user derive sufficient additional benefit from femtocells to value and be prepared to buy into a femtocell service. They also create additional methods for operators to distinguish their femtocell service from those of other competing operators, as femtocells are launched by a greater proportion of operators. As a result they have become of increasing priority for the femtocell community. In Section 11.5 we illustrate the femtozone and connected home service types by reference to a number of examples, while in Section 11.6 we consider the factors which will need to be considered to enable proliferation of such services. Section 11.7 indicates a possible sequence for operators to use for progressively rolling out these services.

11.5 Service Examples

These examples serve to illustrate the service classes and are certainly not intended to be an exhaustive list. Indeed, it is expected that the most interesting and innovative possibilities will only emerge once femtocells have been widely deployed and the service enablers highlighted in the following section have helped application developers to create services for a wide range of femtocells.

11.5.1 Femtozone Services

In some cases, the femtozone service examples in this section could be delivered via mechanisms operating within the home LAN, i.e. as connected home services. In this case, however, all services are delivered directly by the operator.

Figure 11.3 shows how femtocells can enable a range of services based on Web 2.0 applications. When the mobile enters the femtozone, the femtocell detects the presence of the user and, based on preferences which the user has previously expressed, updates a Web 2.0

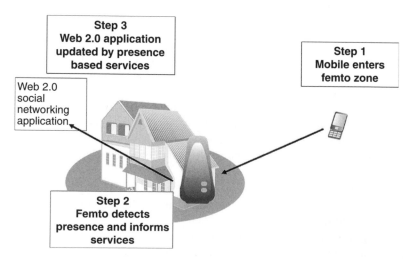

Figure 11.3   Presence and Web 2.0 applications. *Reproduced by permission of Femto Forum Ltd*

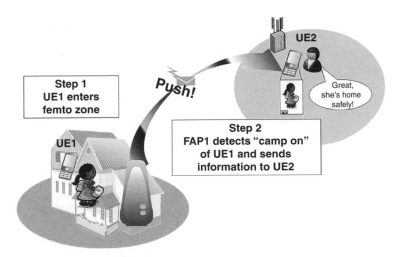

Figure 11.4   'I'm home' service. *Reproduced by permission of Femto Forum Ltd*

application that the user is there. For example, a social networking application can report that a user is in the home, keeping the user's status updated automatically. This information may only be available to a limited set of nominated and trusted friends. The same basic capability could allow a wide range of other services to benefit from the specific knowledge of the user being at home.

In the 'I'm home' service example of Figure 11.4, the femtocell announces the entry of a user into the femtozone to another mobile user, rather than to another application. This would enable, for example, parents to be sure that their children have arrived home safely.

In Figure 11.5, femtocell users can set a number of 'virtual fridge notes', by sending special e-mails or text messages. These messages are delivered to other users when they arrive in the femtozone. They can be acknowledged by the user receiving them and can be set to have

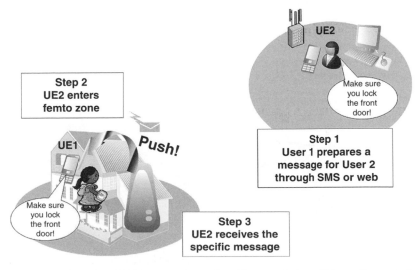

Figure 11.5   Virtual fridge notes. *Reproduced by permission of Femto Forum Ltd*

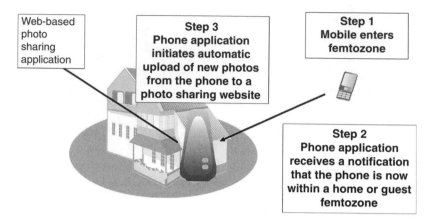

Figure 11.6   Photo upload. *Reproduced by permission of Femto Forum Ltd*

specific lifetimes or to be repeated or flagged according to their importance. Messages are guaranteed to be delivered swiftly on entry to the femtozone.

In Figure 11.6 a camera-phone user enters the femtozone and all of the pictures or videos that have been taken since their last entry to the femtozone are uploaded to the femtocell, which also arranges for the photos to be uploaded to a Web-based photo-sharing application. The pictures can then be automatically deleted from the phone if desired. This ensures that the user avoids having to worry about backing up their photographs, and enables rapid photo sharing without incurring high charges or taking up valuable network resources.

11.5.2 Connected Home Services

Connected home services are distinguished from femtozone services in that they involve specific interaction between the femtocell and devices in the home network other than the mobile devices. In Figure 11.7, the contents of a phone such as the phone book, agenda and

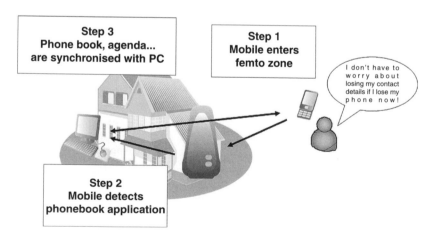

Figure 11.7   Phone book synchronisation. *Reproduced by permission of Femto Forum Ltd*

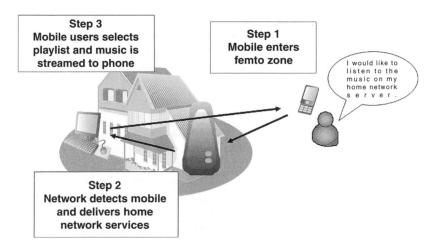

Figure 11.8 Network music server. *Reproduced by permission of Femto Forum Ltd*

contacts are synchronised with a home PC. This synchronisation avoids users worrying that they may lose their valuable contacts and also allows other content such as documents to be updated either on the mobile or on the PC without the user having to remember to copy from one to the other in the correct direction. In this case the services are initiated by the mobile.

Figure 11.8 shows a situation where, on entering the femtozone, the mobile gains access to a variety of devices on the home network which the user can control using their phone. They could, for example, access music stored on a home media server and choose to play that music either on their phone or direct it through their home entertainment system. The same control functions would also be possible when the user is on the macrocell network. The service possibilities of such a personalised universal remote control are endless.

11.6 Service Enablers

The service examples in the previous section are intended to spark the creativity of developers who will doubtless imagine service possibilities far more exciting than those described. In order to develop those services efficiently, however, developers need access to a platform which enables the same applications to work on femtocells from different vendors and with different air-interface technologies. The developer should not need to consider whether the femtocell operates using WCDMA, LTE, CDMA or WiMAX, for example. We can identify several features which are common to many of the services described and where harmonisation or standardisation would be of benefit. Note that, in many cases, this information will be available to the mobile operator via existing mechanisms; the point here is to highlight the need for this information to be available to *applications* and to ensure that the issues raised, such as those regarding points of control and security, are properly addressed.

Identified service enablers include:

- *Presence*: the femtocell needs to provide a consistent means of delivering the identity of the user and of the specific mobile device within the femtozone. It should also identify the type of user, which might be the femto owner, a member of the femtocell main user group, or a guest user. This will enable the application to distinguish between a user being at home and visiting another femtocell user's home for which they have 'visiting rights.'

- *Femtocell capabilities:* the application should gain access to information on the capabilities of the femtocell, such as its identity, its data bearer capabilities and whether it can deliver local IP access.
- *Mobile device information:* the application should gain access to some means of identifying the mobile device and its bearer capability.
- *Status*: the application would often wish to gain access to some information as to the activities that a user is currently engaged in, so that content delivery can be varied accordingly. For example, the streaming of music should be interrupted when the user is engaged in a voice call. The status indication could include whether the current service is circuit or packet switched, whether the call is incoming or outgoing, call duration and/or data volume, and perhaps the current data performance of the device.
- *Service groups*: it should be possible for an application to identify user groups on the femtocell in order to deliver different services for each group. Groups may include individuals, home groups, authorised visitors and general users not granted specific access to a given femtocell but who are nevertheless permitted to gain access for standard services.

11.6.1 Service Implementation

Within the service types and enablers highlighted there is a wide range of possibilities for implementation with their own merits. At the time of writing no single approach has yet emerged and this is likely to be an important debate in future standardisation and harmonisation of femtocells.

In general terms, services and applications can be enabled in the network, in the phone or in the femtocell, in the home gateway, or in a mixture of all four. Network-implemented services help to make applications available to all users and facilitate simple charging and control by operators. Such services require operator APIs to notify Web applications regarding, for example, home presence and may raise issues of performance and scalability in some cases.

Services and service enablers could be implemented directly on phones, which opens the opportunity for many existing applications to be femtocell enabled, but creates challenges of uniform handset APIs on multiple operating systems to allow the applications to differentiate femtocell operation from operation on the macro network.

Applications could operate on the femtocell itself, but this raises issues of potentially limited processing power and memory, plus potential issues with security, application compatibility and software upgrades. On the other hand, it provides a helpful point for operator control and a presence in the home for operators who are not offering an integrated broadband service.

Lastly, applications can reside on the home gateway, making use of existing developments in that sector where many industry organisations have already specified a range of methods for enabling applications. It may, however, limit the rate of take-up of such approaches to the rate of deployment of integrated femto/home gateway solutions and may raise issues of control between the broadband operator and the mobile operator.

11.7 Stages of Femtocell Service Introduction

Services will form a critical component of providing a successful business case to support a femtocell deployment. The question of how to market a femtocell solution successfully is one of the significant questions for the industry today. Marketeers need to come to terms with the unique characteristics of a femtocell solution at the same time as finding a compelling and simple enough marketing message that will resonate with the potential customer base. It is currently emerging that many carriers are taking a staged approach to how they are introducing femtocell solutions into the market. These stages are listed below.

11.7.1 Stage 1 – Supporting Fixed Mobile Substitution

The first stage of the market will really focus upon supporting a current trend. There is a significant swing in usage models for consumers whereby they are pushing more of their total number of communications minutes away from fixed-line solutions to mobile solutions. It is likely that first services to market, and thus the market messaging, will be around fixed–mobile substitution concepts such as 'all-you-can-eat' voice bundles in the home or cheaper pre-pay voice costs.

11.7.2 Stage 2 – Prompting Mobile Data Uptake

The second stage of the service positioning is likely to take the form of supporting and promoting mobile Web-based services that are currently in the market. This could be in the form of general Web browsing to MMS messaging and services. The femtocell solution could be used to reduce the cost in the home environment in order to spur behavioural change and more device usage.

11.7.3 Stage 3 – Bringing the Mobile Phone into the Connected Home Concept

Stage 3 sees the beginning of a more advanced concept that can clearly begin to leverage some of the more advanced capabilities of a femtocell. Items like presence and inherent connectivity can be leveraged to tie the femtocell solution into the connected home. This could involve allowing for data transfer and communication between a mobile device in a good signal environment to items like set-top boxes to home media centres to PCs. This can really allow for a more compelling device environment for the consumer and deliver lifestyle enhancement.

11.7.4 Stage 4 – Taking the Connected Home into the Wider World

The final stage of the femtocell service dynamic is one that is most attractive to the carrier community. This stage involves the enabling of communications with home-based devices with the mobile device in the macro domain. This could take the form of accessing set-top box held visual content or PC-based work files whilst outside of the home. This is the final step of integrating the mobile phone into a true multimedia and communications service bundle and allows the carrier to take advantage of its most pervasive characteristic – wide-area wireless connectivity. It also allows the carrier to generate significant additional revenues by enabling consumers to access content thus driving traffic growth.

11.8 Conclusions

As well as delivering efficient, high-quality access to existing services and operators, femtocells can enable new possibilities for operators, business partnerships and services. These will help to encourage innovation and positive competition in the industry while also creating additional enthusiasm amongst mobile users for gaining access to new and existing services. These possibilities should enable continuity of femtocell innovation for many years to come.

12

Summary: The Status and Future of Femtocells

Simon Saunders

12.1 Summary

The femtocell industry is young, but has already taken significant steps towards maturity. In this book we have attempted to highlight the status of femtocell development and the directions being taken for the future. Here we summarise the key findings of each of the chapters.

Chapter 1: Introduction to Femtocells showed that the continued growth of mobile services, especially for mobile broadband data, creates greater expectation for services, volume, quality and coverage for users and new challenges for operators. Femtocells represent a distinctive new class of device which can deliver economic and service benefits for operators and users alike. They are easy to install and use for the end-customer, while being fully managed by the operator and make full use of the operator's existing spectrum and network assets.

Chapter 2: Small Cell Background and Success Factors explained that small cells are nothing new and indeed have been by far the biggest contributor to the increase in spectrum efficiency of wireless systems over at least the last several decades. For the future, smaller cells are an essential component of wireless networks, delivering solutions to the demand for coverage and capacity at levels which would be uneconomic using macrocells alone. Other small cell types, such as microcells, picocells and distributed antennas, are already in widespread use, but femtocells fill a gap in the options at the smallest scale of homes and offices and are likely also to impact on medium and larger offices. Several commercial and technical factors have combined to make femtocells viable and attractive at the present time in a way which would not have been possible previously.

Chapter 3: Market Issues for Femtocells highlighted the market opportunities and challenges for operators implementing femtocells. Concentrated effort on these factors will assist operators in launching femtocells at the right time to maximally benefit their business and in creating femtocell offers which will be attractive to their customers. Analysis of the business

Femtocells: Opportunities and Challenges for Business and Technology   Simon R. Saunders, Stuart Carlaw, Andrea Giustina,
Ravi Raj Bhat, V. Srinivasa Rao and Rasa Siegberg   © 2009 John Wiley & Sons, Ltd

case for femtocells demonstrated that femtocells can yield a relatively rapid return over a broad range of situations if the correct customer segments are targeted and if the femtocell offer is used to establish cost savings and to drive additional usage of the mobile operator's network. Forecasts suggest rapid growth in femtocells from 2010 onwards given successful negotiation of the technical, standards and market challenges identified.

Chapter 4: Radio Issues for Femtocells surveyed the impact of femtocells on spectrum reuse. It showed that, while interference between femtocells and macrocells is possible in extreme circumstances, a set of mitigation techniques is available to vendors and operators to overcome this in most cases. If these techniques and others are properly implemented by vendors and operators, the network quality is actually improved for both femtocell and macrocell users alike, while the resulting dense frequency reuse delivers a substantial increase in spectrum efficiency and total network capacity. The chapter also highlighted the opportunity to evolve standards for RF performance parameters for femtocells to be appropriate for the radio environments in which they are deployed and explained the factual situation concerning RF health and safety for femtocells.

Chapter 5: Femtocell Networks and Architectures explained the requirements and motivations for evolving femtocell network architectures for femtocells from the traditional architectures for standardised WCDMA, LTE, CDMA and GSM systems. It explained the major options available for such changes and reasons for the particular selections that have already taken place in several standards. It showed that, while further evolution of these standards is expected and necessary in the future, the current standards form a sound basis for initial femtocell deployments and are fully scalable to large volumes. It also highlighted the network issues associated with several important aspects of femtocell operation, including mobility, quality of service, local IP access and femtozone services.

Chapter 6: Femtocell Management explained the operator requirements for configuration and management of femtocells to deliver a zero-touch experience for end-customers while enabling scalable deployments up to millions of femtocells under the operator's full control. It showed how the evolving management data models and procedures have been derived for proven models in the wired broadband markets to deliver against these requirements and explained the status and evolution of current management standards.

Chapter 7: Femtocell Security highlighted the importance of security from both the operator and the customer viewpoint. It provided a model of the major potential security threats and the mechanisms available for addressing these. The potential approaches, rooted in both mobile and Internet standards, were explained, leading to an understanding of the motivations behind the choices that have been made for femtocell security standards.

Chapter 8: Femtocell Standards and Industry Groups highlighted the critical importance of standards to femtocell adoption and evolution. It surveyed the standards that have been published or are in formation across the major mobile standards families, including 3GPP, 3GPP2 and IEEE/WiMAX Forum. It also explained the other industry groups who are working together to encourage and adopt these standards and to ensure implementation, compliance and interworking and to promote awareness of the commercial and service potential for femtocells.

Chapter 9: Femtocell Regulation explained how many regulatory goals are supported by femtocells, spanning technical, market and social factors. It explained that femtocells can fit broadly under existing regulation, but that there are several areas where early evolution and greater clarity in existing regulations would help to encourage and enable operators to deliver these to the benefit of citizens and consumers.

Chapter 10: Femtocell Implementation Considerations examined several of the technical and cost-related factors which make implementation of femtocells challenging. It showed that these factors are all addressable, but require a multi-disciplinary approach, which borrows techniques from a number of fields beyond those associated with traditional base station engineering. The design must also consider system-level cost and technical issues in order to deliver a competitive yet standards-compliant product.

Chapter 11: Business and Service Options for Femtocells highlighted the potential for femtocells to go beyond the efficient delivery of existing business models and services, to creating new possibilities for both operators and mobile users. The network and spectrum assets to deliver a complete femtocell service may lie in the hands of players from several domains, so that a complete offering may encourage operators to form interrelationships, which may challenge traditional models for delivery of mobile services. Femtocells can also constitute a new service-delivery platform, which will encourage the development of brand-new services and add special capabilities to existing services, which would not have been feasible without the femtocell.

In the next section we consider ways in which this background may lead to evolutions of the femtocell concept and applications in the future.

12.2 Potential Future Femtocell Landscape

12.2.1 Growth of Femtocell Adoption

In the early days of femtocells, many associated technical challenges were highlighted, including interference issues, cost, network scalability and management. Careful study, field trials and industry consensus leading to standards, have all contributed to a current consensus that all of these technical issues can be overcome by those with the knowledge and expertise to apply best-practice approaches, many of which have been explained in earlier chapters of this book. Commercial challenges are also significant, including ensuring an appropriately attractive offer to end-customers. At the time of writing, many operators have worked through these challenges to the point that they have been confident enough to launch femtocells and are able to gauge the success of their initial offers and to adjust them over time to maximise their success.

As a result, we expect to see substantial acceleration in the number of launches of femtocells around the world. It once seemed that femtocell deployments would be concentrated on some limited region, but it is now clear that operators in Europe, North and South America, Asia, the Middle East and Australasia are launching femtocells. Beyond initial launch, clearly the rate of uptake of femtocells is uncertain, but given the continued rapid growth in mobile and broadband adoption and of mobile broadband services, the indications are positive. We also expect that once a first operator has launched in a given market, competitive pressures are likely to encourage other operators to respond with a comparable service.

The nature of femtocell offers is also likely to evolve. We could caricature initial femtocell offers as being focused on coverage in the United States, capacity in Europe and new services in Asia. New offers, particularly those which are a competitive response to a first deployment in a given market, are likely to have a more evolved combination of these basic propositions. While most initial offers will focus on delivering existing services, the evolution of new services is likely to be an increasing component of femtocell offers in order to create additional operator differentiation and in response to the adoption of femtocells as a development platform by application developers.

12.2.2 Femtocells in Homes and Offices

Femtocells were initially targeted solely on home environments, where there was no other viable option for delivering true dedicated mobile broadband services. However, operators drove the expansion of this to include offices and in several cases particular operators see office use of femtocells as the driving force for their femtocell launches. Enterprise customers hold considerable influence when negotiating with operators for contract renewal and they may also be prepared to invest in femtocells as part of their IT infrastructure, given an appropriate contract structure. For example, some operators may be prepared to permit enterprises and their IT system integrators to own and operate femtocells in their environments, consistent with certain 'hygiene rules' regarding the spectrum usage, in return for leases on the relevant spectrum and roaming revenue onto the wide-area networks beyond the enterprise.

12.2.3 Femtocells in Developing and Rural Markets

Femtocells are initially targeted at developed markets, where service demand and broadband adoption is greatest and where price points for the first femtocell offerings will deliver a return for operators even where manufacturing volumes are initially fairly low. However, femtocells may also play a role in delivering services in developing markets as a means of enabling mobile uptake by the remaining 40% of the population which does not yet have service access. This will require femtocells which are backhauled in different ways, perhaps via satellite or fixed terrestrial networks. They may also concentrate initially on 2G services to make best use of available low-price handsets. One can imagine, for example, a GSM femtocell, backhauled over a fixed wireless network and with open access, delivering the first mobile voice and data services to rural villages, which are uneconomic to serve in other ways. Such locations can then benefit from the established economic growth potential of access to mobile services. Further, even in developed markets there are often areas of very low density rural population, which would benefit from wirelessly backhauled femtocells both for voice and data applications.

While femtocells targeted at such environments will take some time to develop and establish, the market opportunity and consequent social and economic benefits are potentially immense.

12.2.4 Femtocells Outdoors

Operators have already expressed considerable interest in deploying evolved femtocells into outdoor environments. These are likely to be open access devices deployed on street furniture or outdoor building walls using dedicated but low-cost backhaul (e.g. sDSL) to enhance capacity in areas of dense usage. In essence they would fulfil the same role as microcells, but would do so by building on the economies of scale arising from indoor femtocells and would generate operational cost savings by using the self-managing capabilities first developed for femtocells.

12.2.5 Femtocell-Only Operators

The macrocell has always been the starting point for establishing a new mobile network. However, such services may generate little competitive benefit for a new entrant operator while requiring a large amount of pre-revenue investment. We therefore envisage some new entrants deploying only femtocells, while establishing roaming relationships with wide-area operators. Such a service would require limited spectrum access for those operators, and may even be used by several operators concurrently. Such spectrum could even be shared with other dissimilar services, such as the use of 'white space' spectrum which is shared with digital broadcasting services. Operators of this kind would compete on a combination of price and access to special services.

12.2.6 Femtos Enabling Next-Generation Mobile Networks

Femtocells are expected to play a special role in next-generation mobile networks (NGMN), including LTE, WiMAX and any mobile network technology which qualifies under the ITU-R's definition of IMT-Advanced services (112). The standards for NGMN are all expected to include femtocell features. Operators deploying NGMN will be able to incorporate femtocells into their NGMN strategy with a degree of integration fashion which has not been possible in 3G networks, where femtocells emerged well after initial deployments. User devices will incorporate features which specifically optimise their performance with femtocells and will deliver service capabilities that are specifically created with femtocells in mind.

Femtocells meet particular recommendations of the NGMN Alliance. In particular they fulfil the needs of a 'cost-optimised indoor node design', envisaged as follows (from (113)):

> Cost-optimised indoor node design. In many cases, indoor scenarios will allow a simplified radio modem design, due to simpler propagation situations, reduced MTBF requirements, reduced transmit power etc. However, contemporary design of indoor equipment does not allow an economically viable deployment. The NGMN RAN shall be designed in a way that it allows a large-scale deployment of cost-optimised plug-and-play NGMN-only indoor radio equipment at a price level of commercial quality.

This is virtually a direct definition of a femtocell. Additionally femtocells fulfil the NGMN recommendations for self-organising networks (SON), including the following subtasks:

- self-planning;
- self-configuration;
- self-optimisation and self-tuning;
- self-testing and self-healing.

Indeed, femtocells can be seen as the first commercial instance of a true SON, helping to build confidence and establish specific technical approaches which will be useful for the wider application of SON.

Technically speaking, femtocells are well suited to maximise the benefit associated with common next-generation air-interface features, such as high-order modulation schemes and multiple-input multiple-output (MIMO) antenna schemes, based on spatial processing. The high geometry factors associated with indoor cells (Section 2.2.5) enable the highest order

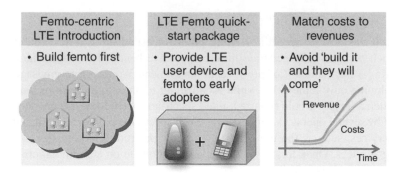

Figure 12.1   New next-generation deployment approaches enabled by femtocells

modulation schemes to operate over almost the whole service area, ensuring that the peak data rate available is also representative of the typical user experience, in contrast to the macrocell environment, where only users close to the macrocell with excellent service quality gain such benefit. The rich three-dimensional multipath environment also provides a radio channel which maximises the 'diversity order' or 'rank' of the channel, enabling MIMO to maximise the data throughput and channel resilience. This advantageous propagation channel also permits the femtocell to incorporate the necessary multiple-element antenna structures for MIMO within very small volumes.

The interference management capabilities of femtocells could also allow spectrum to be used in new ways. For example, next-generation mobile networks will typically support both frequency-division (FDD) and time-division (TDD) modes. Given that user devices will often support both, operators can reuse spectrum which is usually intended for FDD use via femtocells and open up TDD bands which have been less widely used in third-generation systems. They might also be able to use portions of spectrum which were originally intended as guard bands, without creating harmful interference.

Femtocells also enable next-generation mobile networks to benefit from several new deployment approaches, illustrated in Figure 12.1:

- *Femto-centric NGMN deployment:* femtocells can act as the first deployment of NGMN technology. Such deployments will complement rather than replace third-generation networks. Instead of upgrading thousands of macrocells with next-generation before launching a service, the operator can deliver initial NGMN services at an earlier stage and in the home and offices of those with the greatest needs and motivations for NGMN. Beyond these homes and offices multi-mode devices will fall back to third-generation networks, delivering a seamless IP-based user experience.
- *Quick-start packages*: deployment of next-generation services has usually been slowed by the need to get enough user devices capable of next-generation operation into the population of users to make service launch viable. Operators can instead offer early adopters of next-generation technology a package comprising a next-generation femtocell and compatible devices. Such users will benefit from an enhanced user experience, help to road-test and optimise the delivery of new services and can also act as evangelists for the technology to other users, helping to build the user base and the operator business case for more extensive next-generation roll-out.

- *Match costs to revenues*: femtocells can deliver next-generation services to the relevant users and locations with surgical precision. Operators avoid a 'build-it and they will come' approach, where the costs of network upgrade are incurred and user demand fails to emerge as soon and as extensively as envisaged. They can both accelerate and retard roll-out in response to changes in demand with very little delay, in contrast to response times of order 1–3 years with macrocells. Thus costs can be matched very closely to variations in the associated revenues, reducing the financial risks and increasing the overall value associated with next-generation roll-outs. This reduction in risk helps the case for the wider roll-out, even beyond the femtocell element of service.

Finally, the high bit rates available on femtocells in the home and office, combined with the offload capabilities of direct IP access, enable femtocells to be used for local area networking in ways which would not previously have been contemplated. This would allow, for example, the distribution of high-definition video content around home devices, enabling the home or office networking, which has so far been delivered using Wi-Fi, to be combined with the benefits of resilience and QoS associated with licensed spectrum.

12.2.7 When is a Femtocell not a Femtocell?

Throughout this book we have been careful to emphasise that devices only qualify for femtocell status if they fulfil *all* of the attributes specified in Section 1.4. With that caveat, however, we can imagine ways in which femtocell-like technology could be beneficial even if some of those attributes are absent. For example:

- Femtocells could be operated in spectrum which is unlicensed (strictly licence-exempt) or where access is licensed to multiple operators (concurrent access) in order to increase capacity and enable operators who do not have full licensed spectrum access. The use of femtocells in white space spectrum could be one such example.
- The proven self-organising capabilities of femtocells can be incorporated into wide-area cells to deliver operational and technical efficiencies.
- Femtocells could become somewhat mobile devices, operating in locations such as on trains, planes and ships via appropriate wireless backhaul.
- Femtocells deployed in sufficiently high density could be backhauled via links which pass between adjacent femtocells, enabling a mesh topology where locally routed data need not access the fixed-line backhaul and where the limitations of fixed-line availability can be mitigated via distributed backhaul sharing amongst cells.

Other innovative possibilities can be imagined, all enabled by the underlying technology and economies of scale associated with 'conventional' femtocell applications.

12.3 Concluding Remarks

At the time of writing, the femtocell industry had overcome, or was in the process of overcoming, the numerous technical and commercial barriers to initial service launches within a surprisingly short space of time. This was possible because of the strong encouragements from operators to find solutions to pressing issues of mobile coverage, capacity and business growth, and because vendors responded with innovative solutions and cooperated on common challenges. In the time to come, the sharing of best practice learned from the initial technology and feedback from customers should allow operators and vendors to further refine their offerings to overcome any new and unexpected challenges and to enable the scale of femtocell deployments to grow to reach their full potential.

It is hoped that this text will help the process in some small way, by spreading the awareness of femtocells, by encouraging awareness of the challenges and available solutions, and by expanding the community of users and developers who are applying thought to innovating for the next generations of femtocells. If we have missed any important areas we apologise, but we would encourage feedback and suggestions via info@femtocellbook.com and www.femtocellbook.com.

References

1. **GSM Association.** *Mobile data stats.* 2008.
2. **Analysys Mason.** *Wireless network traffic 2008–2015: forecasts and analysis.* [Online] 2009. http:// www.analysysmason.com/Research/Content/Reports/Wireless-network-traffic-20082015-forecasts-and-analysis/.
3. *Internet World Statistics.* [Online] http://www.internetworldstats.com.
4. **Cisco.** *Cisco visual networking index – forecast and methodology, 2007–2012.* [Online] http://www.cisco.com/ cn/US/solutions/collateral/ns341/ns525/ns537/ns705/ns827/white_paper_c11-481360_ns827_Networking_ Solutions_White_Paper.html.
5. **Uni, Finnish.** *Mobile data in Finland.* 2008.
6. **Merrill Lynch.** *Wireless data growth: How far, how fast, and who wins?* 2007.
7. **Mavrakis, Dimitris and Kamal-Saadi, Malik.** *Mobile broadband access at home: the business case for femtocells, UMA and IMS/VCC dual mode solutions.* Informa, 2008.
8. **Northstream.** *UMA paves the way for convergence.* 2005.
9. **VisionGain.** *In-building wireless solutions: stimulating greater mobile usage through better indoor coverage.* 2006.
10. **Ofcom.** *Mobile citizens – adapting regulation for a mobile, wireless world.* London: Ofcom, 2008.
11. **Meredith Sharples, Vodafone UK.** *Femtocell business models for UK operators.* Informa, 2008.
12. **Saunders, Simon.** The future of wireless. three decades of wireless evolution. [book auth.] William Webb. *Wireless communications – future services and technologies.* John Wiley & Sons, Ltd, 2007.
13. **Ring, D.H.** *mobile telephony – wide area coverage.* Bell Telephone Laboratories. http://www.privateline.com/ archive/Ringcellreport1947.pdf, 1947.
14. **Saunders, Simon and Aragon-Zavala, Alejandro.** *Antennas and propagation for wireless communication systems.* 2nd edition. Chichester: John Wiley & Sons, Ltd, 2007.
15. **Aragon-Zavala, Alejandro, Cuevas-Ruiz, Jose and Delgado-Penin, Jose Antonio.** *High-altitude platforms for wireless communications.* Chichester: John Wiley & Sons, Ltd, 2008.
16. **Moore, Gordon.** Cramming more components onto integrated circuits. *Electronics Magazine.* April 1965, Vol. 38, p. 8.
17. **Shannon, C.** A mathematical theory of communication. *Bell System Technical Journal*, 1948, Vol. 27, pp. 379–423 and 623–56.
18. **Jacobs, Paul E.** [Online] http://techon.nikkeibp.co.jp/english/NEWS_EN/20080905/157548/?P=1.
19. **3GPP.** *Requirements for Evolved UTRA (E-UTRA) and Evolved UTRAN (E-UTRAN).* 2008. TR 25.913.
20. **Cooper, Martin.** Cooper's law. [Online] http://www.arraycomm.com/serve.php?page=Cooper.
21. **Rumney, M.** The importance of average vs. peak performance in cellular wireless. *Microwave Journal.* [Online] November 2008. http://www.mwjournal.com/news/article.asp?HH_ID=AR_6833.
22. **Holma, H. and Toskala, A.** *WCDMA for UMTS.* 3rd edition. Chichester: John Wiley & Sons, Ltd, 2004.
23. **NGMN Alliance.** *NGMN optimised backhaul requirements.* [Online] August 2008. http://www.ngmn.org.
24. **Tolstrup, M.** *Indoor radio planning: a practical guide for GSM, DCS, UMTS and HSPA.* Chichester: John Wiley & Sons, Ltd, 2008.

25. **IEEE.** *IEEE LAN/MAN wireless LAN standards.* [Online] http://standards.ieee.org/getieee802/802.11.html.

26. **Wi-Fi Aliance.** *Wi-Fi Alliance website.* [Online] http://www.wi-fi.org/.

27. **Organisation for Economic Cooperation and Development.** OECD broadband portal. [Online] http://www.oecd.org/sti/ict/broadband.

28. **Ovum.** *Emerging markets and data will drive global mobile growth to 2013.* [Online] http://www.ovum.com/news/euronews.asp?id=7584.

29. **GSMA.** GSMA market data summary. [Online] http://www.gsmworld.com/newsroom/market-data/market_data_summary.htm.

30. **M:Metrics.** comScore M:Metrics reports mobile search grew 68 percent in the U.S. and 38 percent in Western Europe during past year. *M:Metrics.* [Online] September 2008. http://www.comscore.com/press/release.asp?press=2469.

31. **M:Metrics.** Americans spend more than 4.5 hours per month browsing on smartphones, nearly double the rate of the British. *M:Metrics.* [Online] May 2008. http://www.mmetrics.com/press/PressRelease.aspx?article=20080521-smartbrowsing.

32. **Ofcom.** *Communications market report.* 2008.

33. **Baines, R.** The picoArray: A reconfigurable SDR processor for basestations. [book auth.] W. Tuttlebee. *Software defined radio: baseband technologies for 3G handsets and basestations.* Chichester: John Wiley & Sons, Ltd, 2003.

34. **Moore, Gordon.** Excerpts from a conversation with Gordon Moore. [Online] 2005. ftp://download.intel.com/museum/Moores_Law/Video-Transcripts/Excepts_A_Conversation_with_Gordon_Moore.pdf.

35. **Gao, X.** Effective Moore's laws in high performance computing.

36. **Carlaw, S and Wheelock, C.** *Femtocells market challenges and opportunities – cellular based fixed mobile convergence for consumers, enterprises and SMEs dedicated to benefits of femtocells.* ABI Research, 2007.

37. **Fong, Serene.** *DSL market forecasts. MD-BSUB-101.* ABI Research. 2008.

38. **Lee, Paul.** *Fibre to the premises market forecasts. MD-FTTX-102.* ABI Research. 2008.

39. **Carlaw, Stuart.** *The role of home access points in the drive for fixed mobile convergence.* Dallas: Avren Events, 2007. Femtocells Europe.

40. **ABI Research.** *Consumer wireless survey.* 2008.

41. **Carlaw, Stuart.** *Femtocell business models. RR-FBM-08.* ABI Research. 2008.

42. **Carlaw, Stuart.** *Challenges facing the Femtocell Market - A Realistic View?* Dallas: Avren Events, 2007. Second International Conference on Home Access Points and Femtocells.

43. **International Telecommunication Union.** *Propagation data and prediction models for the planning of indoor radio communication systems and radio local area networks in the frequency range 900 MHz to 100 GHz.* Geneva: ITU-R, 1997. ITU-R Recommendation P.1238.

44. **3GPP.** *FDD base station (BS) classification.* 2008. TR 25.951.

45. **Ericsson.** *Home NodeB output power.* 3GPP TSG-RAN WG4, 2007. R4-070969.

46. **3GPP.** *3G Home NodeB study item technical report.* 3GPP, 2008. TR25.820.

47. **Femto Forum.** *Interference management in UMTS femtocells.* www.femtoforum.org, 2008.

48. **3GPP.** *FDD Home NodeB RF requirements.* TR 25.967.

49. **Will Franks, Ubiquisys.** *Femtocell practical: lessons learned from real world deployments.* Dallas: Avren Events, December, 2007. Second International Conference on Home Access Points and Femtocells.

50. **3GPP.** *Base station (BS) radio transmission and reception (FDD).* 2008. TS 25.104.

51. **3GPP.** *Base station conformance testing (FDD).* 2008. TS 25.141.

52. **Repacholi, M. H. and Cardis, E.** Criteria for EMF health risk assessment. *Radiat Prot Dosim*, 1997, Vol. 72, pp. 305–312.

53. **Fujimoto, K. and James, J.R.** *Mobile antenna systems handbook.* 2nd edition. London: Artech House, 2000.

54. **IEEE SCC34/SC2.** *IEEE recommended practice for determining the spatial-peak specific absorption rate (SAR) in the human body due to wireless communications devices: experimental techniques.* 2000.

55. **TC211 WGMBS European Committee for Electrotechnical Standardisation.** *Basic standard for the measurement of specific absorption rate related to human exposure to electromagnetic fields from mobile phones (300MHz–3GHz).* 2000.

56. **CENELEC.** *Basic standard for the calculation and measurement of electromagnetic field strength and SAR related to human exposure from radio base stations and fixed terminal stations for wireless telecommunication systems (110MHz–40GHz).* European Standard EN50383.

57. **CENELEC.** *Product standard to demonstrate the compliance of radio base stations and fixed terminal stations for wireless telecommunication systems with the basic restrictions or the reference levels related to general public exposure to radio frequency electromagnetic fields (110MHz–40GHz).* 2001. European Standard EN50385.

58. **CENELEC.** *Generic standard to demonstrate the compliance of low power electronic and electrical apparatus with the basic restrictions related to human exposure to electromagnetic fields (10MHz–300GHz) – general public.* European Standard EN 50371.

59. **CENELEC.** *Basic standard for the measurement of Specific Absorption Rate related to human exposure to electromagnetic fields from mobile phones (300MHz–3GHz).* 2001. European Standard EN50361.

60. **International Commission on Non-Ionizing Radiation Protection.** Guidelines for limiting exposure to time-varying electric, magnetic, and electromagnetic fields (up to 300GHz). *Health Physics*, 1998, Vol. **75**, pp. 494–522.

61. **Standardisation, TC211 WGMBS European Committee for Electrotechnical.** *Basic standard for the measurement of specific absorption rate related to human exposure to electromagnetic fields from mobile phones (300 MHz–3GHz).* 2000.

62. **IEEE.** *IEEE standard for safety levels with respect to human exposure to radio frequency electromagnetic fields, 3 kHz to 300 GHz.* 1991. IEEE C95.1-1991.

63. **Femto Forum, Mobile Manufacturers Forum & GSM Association.** *Femtocells and health.* [Online] www.femtoforum.org. 2007.

64. **IETF.** *Internet Key Exchange (IKEv2) protocol.* December 2005. IETF RFC 4306.

65. **IETF.** *Internet X.509 public key infrastructure certificate and Certificate Revocation List (CRL) profile.* April 2002. RFC 3280.

66. **IETF.** *IP Encapsulating Security Payload (ESP).* November 1998. RFC 2406.

67. **IETF.** *IP authentication header.* December 2005. RFC 4302.

68. **3GPP.** *Network architecture (Release 5).* 2004. TS 23.002 V5.12.0.

69. **3GPP.** *UTRAN architecture for 3G Home NodeB; Stage 2 (Release 8).* 2008. TS 25.467 V8.0.0.

70. **3GPP.** *UTRAN Iuh interface RANAP User Adaption (RUA) signalling (Release 8).* 2008. TS 25.468 V8.0.0.

71. **3GPP.** *UTRAN Iuh interface Home NodeB Application Part (HNBAP) signalling (Release 8).* 2008. TS 25.469 V8.0.0.

72. **3GPP.** *IP Multimedia Subsystem (IMS); Stage 2 (Release 5).* 2006. TS 23.228 V5.15.0.

73. **3GPP.** *IP Multimedia Subsystem (IMS) centralized services (Release 8).* 2008. TR 23.892 V8.0.1.

74. **IETF.** *Definition of the Differentiated Services field (DS Field) in the IPv4 and IPv6 headers.* December 1998. RFC 2474.

75. **3GPP.** *Location management procedures (Release 5).* 2003. TS 23.012 V5.2.0.

76. **3GPP.** *UTRAN Iu interface: general aspects and principles (Release 5).* 2004. TS 25.410 V5.4.0.

77. **Broadband Forum.** *CPE WAN management protocol v1.1.* December 2007. TR-069 Amendment 2.

78. **3GPP2.** *Network architecture model for cdma2000 femtocell enabled systems.* S.P0135-0.

79. **WiMAX Forum.** *Requirements for WiMAX femtocell systems.* 2009.

80. **3GPP.** *Digital cellular telecommunications system (Phase 2+); network architecture (GSM 03.02 version 7.1.0 Release 1998).* 2000. TS 100 522 V7.1.0.

81. **3GPP.** *Digital cellular telecommunications system (Phase 2+); General Packet Radio Service (GPRS); service description; Stage 2 (Release 1998).* 2002. TS 03.60 V7.9.0 .

82. **3GPP.** *Evolved Universal Terrestrial Radio Access Network (E-UTRAN); architecture description (Release 8).* 2008. TS 36.401 V8.4.0 .

83. **3GPP.** *Service requirements for Home NodeBs and Home eNodeBs (Release 9).* 2009. TS 22.220.

84. **3GPP.** *Security of H(e)NB.* TR 33.820.

85. **3GPP.** *Telecommunications management; Home NodeB (HNB) Operations, Administration, Maintenance and Provisioning (OAM&P); Concepts and requirements for Type 1 interface HNB to HNB Management System (HMS).* TS32.581.

86. **3GPP.** *Telecommunications management; Home NodeB (HNB) Operations, Administration, Maintenance and Provisioning (OAM&P); Information model for Type 1 interface HNB to HNB Management System (HMS).* TS32.582.

87. **3GPP.** *Telecommunications management; Home NodeB (HNB) Operations, Administration, Maintenance and Provisioning (OAM&P); Procedure flows for Type 1 interface HNB to HNB Management System (HMS).* TS32.583.

88. **3GPP.** *Telecommunications management; Home NodeB (HNB) Operations, Administration, Maintenance and Provisioning (OAM&P); XML definitions for Type 1 interface HNB to HNB Management System (HMS).* TS32.584.

89. **3GPP2.** *System requirements for femtocell systems.* 2008. S.R0126-0.

90. **3GPP2.** *Femto interoperability specifications.* 2009. A.S0024.

91. **3GPP2.** *Cdma2000 1x specification.* C.S0000x-E.

92. **3GPP2.** *EV-DO standard.* C.S0024-C.

93. **3GPP2.** *Enhanced System Selection (ESS).* 2009. C.S0016-D.

94. **IEEE 802.16.** *IEEE 802.16m system requirements.* 2008. IEEE 802.16m.

95. **GSMA.** *Femtocell interference and frequency.* 2008. GSMA FCG.02.

96. **GSMA.** *Security issues in femtocell deployment.* GSMA FCG.05.

97. **GSMA.** *Management of femtocells.* 2008. GSMA FCG.04.

98. **GSMA.** *Femtocell requirements on DSL broadband.* 2008. GSMA FCG.01.

99. **GSMA.** *3G Femto solution implementation guidelines.* 2008. GSMA FCG.03.

100. **OECD.** *OECD report on links between broadband growth and economic growth.*

101. **European Commission Renewable Energy Unit.** *Code of conduct on energy consumption of broadband equipment Version 3.* 2008. http://re.jrc.ec.europa.eu/energyefficiency.

102. **FCC.** *Enhanced 9-1-1 – wireless services. FCC.* [Online] 2008. http://www.fcc.gov/pshs/services/911-services/enhanced911/.

103. **European Council.** Lawful interception of telecommunications. 17 January 1995. *Official Journal* C 329.

104. **US Government.** Communications Assistance for Law Enforcement Act. *Pub. L. No. 103-414, 108 Stat. 4279, codified at 47 USC 1001-1010.* 1994.

105. **Convention on Cybercrime.** Budapest, 23 November 2001.

106. **ETSI.** *Lawful intercept.* [Online] http://portal.etsi.org/li/Summary.asp.

107. **BABT.** *R&TTE Directive.* [Online] 2009. http://www.babt.com/rtte-directive.asp.

108. **Radio Spectrum Committee.** *Regulatory aspects of femtocells.* Brussels: European Commission, 2008. RSCOM(08)40.

109. **National Institue of Standards and Technology.** *NIST IEEE 1588 website.* [Online] http://ieee1588.nist.gov/. IEEE 1588-2008.

110. **The NTP Public Services Project.** [Online] http://support.ntp.org.

111. **Bubley, Dean.** *Femtocell-aware mobile handsets.* London: Disruptive Analysis Ltd, 2008.

112. **ITU-R.** *Requirements, evaluation criteria and submission templates for the development of IMT-Advanced.* 2008. Report M.2133.

113. **NGMN Alliance.** *NGMN White Paper: next generation mobile networks beyond HSPA & EVDO.* NGMN Allianced, 2006.

114. **Meredith Sharples, Vodafone UK.** *Femtocell business models for UK operators.* Informa, 2008.

115. **3GPP.** *3G Security; Network Domain Security; IP network layer security.* 2008. TS 33.210 V8.1.0.

116. **3GPP.** *MAP application layer security (Release 7).* TS 33.200.

117. **3GPP.** *3G security; Wireless Local Area Network (WLAN) interworking security.* TS 33.234.

118. **IETF.** *Multiple authentication exchanges in the Internet Key Exchange (IKEv2) protocol.* 2006. IETF RFC 4739.

119. **OMTP.** *Security threats on embedded consumer devices.* 22 May 2008.

120. **OMTP.** *Advanced trusted environment OMTP TR1.* 22 May 2008.

121. **Broadband Forum.** *Data model template for TR-069-enabled devices.* 2006. TR-106 Amendment 1.

122. **Broadband Forum.** *TR-196: Femto Access Point Service Data Model,* [Online] 2009. http://www.broadband-forum.org/.

123. **3GPP2.** *Femto security framework.* S.P0132-0.

124. **ITU-T.** *TMN (telecommunications management network) management functions.* 2000. ITU-T Recommendation M. 3400.

125. **Broadband Forum.** *Internet gateway device data model for TR-069.* 2008. TR-098 Amendment 2.

126. **Broadband Forum.** *Auto-config: architecture and framework.* 2002. TR-046.

127. **W3C.** *Simple object access protocol (SOAP) 1.1.* [Online] http://www.w3.org/TR/2000/NOTE-SOAP-20000508/.

128. **3GPP.** *3G security; network domain security (NDS); IP network layer security.* TS 33.210.

129. **3GPP.** *MAP application layer security.* TS 33.200.

130. **3GPP.** *3G security; wireless local area network (WLAN) interworking security.* TS 33.234.

131. **3GPP2.** *Femto security framework.* S.R0132-0.
132. **Wikipedia.** Syzygy. *Wikipedia.* [Online] http://en.wikipedia.org/wiki/Syzygy.
133. **Ravi Raj Bhat and Rao, V. Srinvasa.** *CPE WAN Management Protocol and Femto Access Point Management RF DesignLine, a TechOnline community.* [Online] 2008. www.rfdesignline.com.
134. **Broadband Forum.** Femto access point service data model TR-196, 2009.
135. **3GPP2.** *Femto network overview and list of parts.* X.P0059-000.
136. **3GPP2.** *1xRTT femto access point service specification.* X.P0059-200.

Further Reading

The following websites provide frequently updated sources of information in the fast-moving field of femtocells (links to these and more are available at www.femtocellbook.com):

- www.femtoforum.org – the website of the Femto Forum.
- www.3ginthehome.com – a weekly roundup of femtocell news produced by femtocell vendor ip.access
- www.femtohub.com – an extensive archive of femtocell news, maintained by femtocell vendor Airvana.
- www.thinkfemtocell.com – an independent blog covering many aspects of femtocells.
- www.3gpp.org – the website of the 3G partnership programme, including all of the standards for GSM, WCDMA, HSDPA and LTE. Standards for WCDMA and LTE femtocells are freely available, termed Home NodeB and Home eNodeB, respectively.
- www.3gpp2.org – the website of the 3G partnership programme 2, covering CDMA standards including cdma2000, 1xEV-DO, 1xEV-DO Rev A and UMB.
- www.wimaxforum.org – the WiMAX Forum.

See also the websites of the many femtocell vendors listed at:

- www.femtoforum.org/femto/membership.php

A number of analysts have also produced useful reports on femtocells, including:

- ABI Research – www.abiresearch.com/products/service/Femtocells_Research_Service
- Analysys Mason – www.analysysmason.com/Research/Publications/Reports/Mobile/
- Disruptive Analysis – www.disruptive-analysis.com/femto-aware_handsets.htm
- Informa Telecoms & Media – http://www.informatm.com/access

Femtocells: Opportunities and Challenges for Business and Technology   Simon R. Saunders, Stuart Carlaw, Andrea Giustina,
Ravi Raj Bhat, V. Srinivasa Rao and Rasa Siegberg   © 2009 John Wiley & Sons, Ltd

Appendix

A Brief Guide to Units and Spectrum

A variety of units are used in this book, particularly as regards quantities and speeds of data. There are also references to various spectrum bands. This appendix is intended to assist those unfamiliar with these terms.

The SI system of units is adopted for small-cell terminology and for describing large quantities of data. The original SI prefixes are listed in Table A.1. To complicate matters, when describing quantities of binary data, the prefix may indicate a multiple of either 1000 or 1024 bytes, so we have the outcomes listed in Table A.2. There is some tendency to use 1000 for data rates (so 1 kbps = 1000 bits per second) and 1024 for data quantities (so 1 kB = 1024 bytes) but there is no consistent rule.

There is less ambiguity in the description of frequencies, where 1 MHz = 1,000,000 hertz and 1 GHz = 1,000 MHz. Frequencies are described in general terms as VHF (30–300 MHz), UHF (0.3–3.0 GHz) or SHF (3–30 GHz). Regarding specific frequency bands, femtocells can be deployed in any of the frequency bands supported by the relevant mobile standard, although initial deployments favour the use of the higher frequency bands such as the 1.9 GHz or 2.1 GHz bands. The main mobile frequency bands in current use around the world are shown in Table A.3.

Table A.1 Prefixes used in SI units

Multiples	Name	deca-	hecto-	kilo-	mega-	giga-	tera-	peta-	exa-	zetta-	yotta-
	Symbol	da	h	k	M	G	T	P	E	Z	Y
	Factor	10^1	10^2	10^3	10^6	10^9	10^{12}	10^{15}	10^{18}	10^{21}	10^{24}

Subdivisions	Name	deci-	centi-	milli-	micro-	nano-	pico-	**femto-**	atto-	zepto-	yocto-
	Symbol	d	c	m	μ	n	p	**f**	a	z	y
	Factor	10^{-1}	10^{-2}	10^{-3}	10^{-6}	10^{-9}	10^{-12}	$\mathbf{10^{-15}}$	10^{-18}	10^{-21}	10^{-24}

Femtocells: Opportunities and Challenges for Business and Technology   Simon R. Saunders, Stuart Carlaw, Andrea Giustina, Ravi Raj Bhat, V. Srinivasa Rao and Rasa Siegberg   © 2009 John Wiley & Sons, Ltd

Table A.2 Data quantities

Name	Symbol	Quantity	Alternative
kilobyte	kB	1,024 bytes	1,000 bytes
megabyte	MB	1,048,576 bytes	1,000,000 bytes
gigabyte	GB	1,073,741,824 bytes	1,000,000,000 bytes
terabyte	TB	1,024 gigabytes	1,000 gigabytes
petabyte	PB	1,024 terabytes	1,000 terabytes

Table A.3 International mobile frequency bands

Uplink (mobile transmit, base receive)	Downlink (base transmit, mobile receive)	Duplex mode
777–787 MHz	746–756 MHz	FDD
788–798 MHz	758–768 MHz	FDD
824–849 MHz	869–894 MHz	FDD
830–840 MHz	875–885 MHz	FDD
880–915 MHz	925–960 MHz	FDD
1427.9–1452.9 MHz	1475.9–1500.9 MHz	FDD
1710–1785 MHz	1805–1880 MHz	FDD
1710–1755 MHz	2110–2155 MHz	FDD
1710–1770 MHz	2110–2170 MHz	FDD
1749.9–1784.9 MHz	1844.9–1879.9 MHz	FDD
1850–1910 MHz	1930–1990 MHz	FDD
1850–1910 MHz	1850–1910 MHz	TDD
1880–1920 MHz	1880–1920 MHz	TDD
1900–1920 MHz	1900–1920 MHz	TDD
1910–1930 MHz	1910–1930 MHz	TDD
1920–1980 MHz	2110–2170 MHz	FDD
1930–1990 MHz	1930–1990 MHz	TDD
2010–2025 MHz	2010–2025 MHz	TDD
2300–2400 MHz	2300–2400 MHz	TDD
2500–2570 MHz	2620–2690 MHz	FDD
2570–2620 MHz	2570–2620 MHz	TDD

Index